Argeu Cavalcante Fernandes
Fábio Garcia Penha

The inclusion of nanotechnology in chemistry teaching

Argeu Cavalcante Fernandes
Fábio Garcia Penha

The inclusion of nanotechnology in chemistry teaching

Investigative activities in focus

ScienciaScripts

Cover image: www.ingimage.com

This book is a translation from the original published under ISBN 978-3-330-99675-5.

Publisher:
Sciencia Scripts
is a trademark of
Dodo Books Indian Ocean Ltd. and OmniScriptum S.R.L publishing group

120 High Road, East Finchley, London, N2 9ED, United Kingdom
Str. Armeneasca 28/1, office 1, Chisinau MD-2012, Republic of Moldova, Europe
Managing Directors: Ieva Konstantinova, Victoria Ursu
info@omniscriptum.com

Printed at: see last page
ISBN: 978-620-8-60255-0

SUMMARY

PRESENTATION

According to the guidelines of the National Curriculum Parameters for Secondary Education in the area of natural sciences (PCNEM) and the National Curriculum Guidelines for Basic Education (DCNEB), technological advances and scientific development should be worked on in schools. Nanotechnology is a multidisciplinary area of knowledge and should be taught in a contextualised and interdisciplinary way as a result of this scientific and technological development in contemporary society.

With this in mind, this book is based on the assumption that nanotechnology should contribute to students' scientific education, but that it is rarely dealt with by chemistry teachers in secondary schools. Often, work on this subject is limited to research requested of the students by the teacher. In other words, there is no proposal to link science curricula to nanotechnology and research into teaching chemistry in this area can still be considered incipient.

The research is a qualitative case study and took place in a sixth-period class on the Full Degree in Chemistry course at the IFRN Pau dos Ferros campus during the Supervised Internship Seminar subject. The research showed that it is possible to work on nanotechnology in an investigative way, using methodological tools and strategies that are already known in chemistry teaching. The challenges for the initial training of chemistry teachers certainly include the need to educate undergraduates about new themes that discuss the development of science and technology, using more comprehensive curricula that include teaching nanotechnology, for example, in their didactic-pedagogical proposals.

It was possible to observe that most of the subjects surveyed were interested in the topic, but felt insecure about working on it during their internship. The reasons for this could lie in the insecurity common to trainees, a lack of mastery of the subject because they don't know it well, or even the resistance of the subject teachers to breaking the sequence of content in the curriculum adopted by the school.

The book is structured in three chapters, in addition to the introduction, which presents the problem, the delimitation of the topic, the justification and the proposed objectives. The representative aspects raised in this initial chapter will be taken up in more theoretical and methodological depth in the following chapters.

Chapter 2 presents the theoretical framework on nanotechnology and chemistry teaching. In this chapter, the theme is approached as the fruit of advances in modern science, contributing to a solid scientific education; to this end, its applications, implications and prospects are reported. With regard to the teaching of chemistry, it is seen from a historical perspective without much scientific tradition, given that science teaching in Brazil historically refers to the scholastic and literary model, inherited from the Jesuits. Finally, some methodological strategies are proposed for use in teaching nanotechnology.

Chapter 3 describes the methodological procedures that guided the research, highlighting the research approach, the research subjects, the data collection instruments, the description of the activities carried out and the data analysis methodology.

Finally, Chapter 4 analyses and discusses the results of the research, highlighting the contributions and conclusions of the work for future research in chemistry teaching. It emphasises the need to more effectively include themes related to the development of new technologies, as proposed by the PCNEM and DCNEB, in the curricula of initial training courses for chemistry teachers.

CHAPTER 1

INTRODUCTION

According to the guidelines of the National Curriculum Parameters for Secondary Education in the area of natural sciences (PCNEM) and the National Curriculum Guidelines for Basic Education (DCNEB), technological advances and growing scientific development should be worked on in schools with themes such as nanotechnology. Like the other sciences, chemistry opens up horizons and contributes to the development of this technology, so chemistry lessons are favourable for debate and discussion on the subject.

Within the school context there are relevant aspects that contribute to the critical-reflexive strengthening of students' perceptions of important issues that concern society and the environment. Nanotechnology is a multidisciplinary area of knowledge and should be taught in a contextualised and interdisciplinary way as a result of scientific and technological developments in contemporary society.

Recent scientific research shows that the teaching of nanotechnology in basic education is restricted to activities using videos/documentaries, debates and reading science texts (PAULINI DE JESUS and HIGA, 2014). However, society in general is brought directly or indirectly into contact with advances in nanotechnology, often without having any scientific knowledge, as it only circulates in academic and industrial production circles.

With this in mind, this book is based on the assumption that nanotechnology should contribute to students' scientific and technological education, but that it is rarely dealt with by high school chemistry teachers. Often, work on this subject is limited to research requested of the students by the teacher.

In other words, there is no proposal to link science curricula to nanotechnology and research into teaching chemistry in this area can still be considered incipient. Part of this problem can be found in initial teacher training courses, whose curricula do not include teaching the applications and implications of nanotechnology for society and the environment.

Teachers may not yet feel encouraged to break away from the traditional sequence of teaching methods, which are restricted to chemistry content that has been traditionally established over the years. As well as realising the importance of this topic in chemistry teaching, these teachers need to be encouraged to work with their students, from an investigative perspective, on the evolution of knowledge and its implications for modern man.

In this context, this book is based on the following guiding question: what theoretical and practical didactic parameters can be developed in initial training courses for chemistry teachers, to enable undergraduates to plan and implement the theme of nanotechnology in chemistry classes, in a multi- and interdisciplinary way?

To do this, we started by analysing the representations of the subject of nanotechnology during the investigative observations of the research in a sixth-period chemistry degree course at the Federal Institute of Rio Grande do Norte (IFRN) Pau dos Ferros campus.

The work with the class gave rise to proposals for methodological strategies that could help future chemistry teachers to teach nanotechnology, inciting discussion about initial teacher training with a view to mobilising them to diversify their practices with themes related to the production of new technologies and scientific development.

These methodological strategies, such as investigative experimental activities; teaching mediated by illustrative materials such as documentaries, comic strips and nano-art, show a teaching model that is not in line with that proposed in traditional teaching, as they propose greater interaction between the student and knowledge, with the student himself being an active subject in the construction of his learning.

The general aim of the research is to investigate how nanotechnology can be introduced during the initial training of chemistry teachers through investigative activities. The following specific objectives are related to the general objective:

- To analyse whether nanotechnology is being addressed during the training of future chemistry teachers;
- Proposing investigative activities as methodological strategies for teaching nanotechnology in secondary schools;
- To investigate how trainee teachers view the teaching of current topics such as nanotechnology;
- Relate the knowledge of nanotechnology acquired in secondary school to scientific and technological development;
- Associating the contents of chemistry with the teaching of nanotechnology;

CHAPTER 2

THEORETICAL BACKGROUND

This chapter deals with the bibliographical review relating to the development of the dissertation, comparing the theoretical bases of other works and authors on chemistry education, scientific-technological development and nanotechnology. To this end, it was necessary to resort to scientific articles, essays and books dealing with the three axes mentioned above in order to dialogue and support the analysis and discussion of the results.

The chapter is subdivided into three sections, namely: nanotechnology and chemical education, the teaching of chemistry in Brazil in its historical context and methodological strategies used in the teaching of nanotechnology.

The first section discusses concepts and definitions of nanotechnology, scientific and technological development, as well as the contributions of science to the growth of modern society. It also discusses the applications and implications of nanotechnology for humans and the environment, showing future prospects based on scientific ethics. Chemical education is shown to be essential for understanding the world and for the conscious use of science and technology, and therefore deserves to be highlighted in science education research.

The second section gives a historical overview of chemistry teaching in Brazil and shows that it still has little scientific tradition. Great emphasis is placed on the means of publicising and promoting research carried out in chemistry teaching to show the importance of the partnerships that have been working. It also discusses the need for chemistry teacher training, which has long been neglected by the state.

The last section brings the applicability of all this study to the classroom, so that teachers have the necessary support for their teaching practice in nanotechnology. To this end, some methodological strategies are outlined, such as the use of investigative activities, illustrative teaching materials and the construction of concept maps.

2.1 NANOTECHNOLOGY AND CHEMISTRY EDUCATION

In a society where scientific development is ever increasing, there is a growing need to delve deeper into science in the search for explanations in the depths of matter, in other words, in atoms and molecules. The part of science related to this study is nanotechnology, which has been growing and taking up space in the media thanks to discoveries of nanomaterials in the areas of agriculture, medicine, energy, electronics, computing and others. Its study concerns the manipulation of materials on an atomic scale, from atoms and molecules whose scales are of the order of one billionth of metre, 10^{-9}m (RIBOLDI, 2009).

Many scientific works point to Richard Feynman as the "prophetic" forerunner of nanotechnology knowledge because of his 1959 lecture at the American Physical Society entitled "There's plenty of empty space down there". On this occasion, Feynman provoked scepticism among his listeners when he claimed that it was possible to manipulate atoms and molecules like bricks in a wall (ALVES FILHO, 2008; MEDEIROS and MATTOSO, 2006).

However, for Schulz (2013), there are historiographical distortions regarding the merits attributed to Feynman, as other researchers had already been publishing scientific papers on nanomaterials. These papers dealt mainly with the science of colloids and molecular engineering, such as silver nanoparticles, which were patented in 1954, and the article Molecular Engineering by physicist Arthur Von Hippel, published in 1956, which included all the fundamentals of nanotechnology.

Medeiros and Mattoso (2006) emphasise the importance of creating instruments that would allow man to manipulate matter in nanomanipulations. But it wasn't until the early 1980s that this process was made possible with the help of scanning probe microscopes (SPM), scanning tunnelling microscopes (STM), near field microscopes (NFM) and atomic force microscopes (AFM). These instruments have become the "fingers" and "eyes" of researchers in nanometre-scale manipulation.

According to Quina (2004), nanotechnology works with particles and interfaces on extremely small scales, ranging from 1 to 100 nanometres[I] . For Medeiros and Mattoso (2006, p. 20), it "concerns materials and systems whose structures and components exhibit significantly new and modified physical, chemical and/or biological properties and phenomena due to their nanometric scale".

Particles with nonmetric sizes have a large surface area, and most of them have mechanical, optical, magnetic or chemical properties that are different from those found on macroscopic surfaces. The study and application of the properties mentioned above, which are characteristic of nanomaterials, form the basis of scientific knowledge of this new technology that has emerged with expectations of revolutionising the scientific world in the current century.

This can be measured by observing that the development of so-called Nanosciences and Nanotechnologies (N&N) is taking on ever greater proportions in world-leading countries such as the United States of America, Japan and the countries of the European Community, which are investing in initiatives to provide competitiveness between companies in the technology sector. This incentive is based on the improvements that N&N can offer society, such as: agriculture, energy, quality of life, environmental preservation, public health and many others (ABDI, 2008).

[I] A nanometre is equivalent to a billionth of a metre, i.e. 10^{-9}m. It's like comparing the size of a marble to the size of the planet Earth. On this scale of such small sizes, materials have marvellously new properties that stir man's imagination and challenge scientists and researchers in the 21st century.

Among these new technologies, carbon nanotubes are a good example of some of the advances in this area of science. Studies into the formation of carbon nanotubes were first carried out by Lijima in 1991. Lijima managed to produce carbon nanotubes by pyrolysis of graphite (burning in the absence of oxygen) in a gaseous atmosphere of helium (HERBST, et al, 2004). Carbon nanotubes have great potential to replace silicon in the manufacture of chips and memory cards due to their high storage capacity and lower cost (HERBST, et al, 2004).

However, the techniques most commonly used today to produce carbon nanotubes, both single-walled and multi-walled, are laser ablation and arc discharge (figure 01). In laser ablation, a graphite rod is vaporised by a laser inside a furnace at 1200 °C in an inert atmosphere of argon or helium at a pressure of 500 Torr. In the arc discharge technique, an electrical discharge is passed between two electrodes inside an inert gas chamber. Both techniques produce carbon nanotubes in large quantities, but the arc discharge technique produces an appreciable number of carbon structures and amorphous carbon that requires an additional purification step to obtain the carbon nanotubes (FERNANDES, 2008).

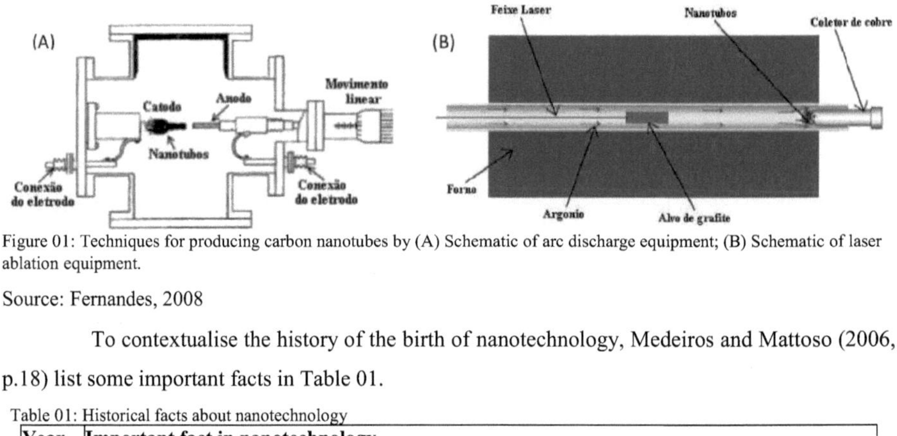

Figure 01: Techniques for producing carbon nanotubes by (A) Schematic of arc discharge equipment; (B) Schematic of laser ablation equipment.

Source: Fernandes, 2008

To contextualise the history of the birth of nanotechnology, Medeiros and Mattoso (2006, p.18) list some important facts in Table 01.

Table 01: Historical facts about nanotechnology

Year	Important fact in nanotechnology
1959	Richard Feynman gave the lecture *a long time ago.* In his speech he proposed atom-by-atom manipulation. It wasn't until two decades later that Feynman's ideas were actually realised.
1974	Researcher Norio Taniguchi gives the name nanotechnology to the field of engineering on a sub-micrometre scale.
1981	Researchers Gerd Binning and Henrich Rohrer (IBM) invented the scanning tunnelling microscope (STM).
1981	The first scientific article on nanotechnology was published by researcher K. Eric Drexler.
1996	Richard Smalley develops a method for producing nanotubes with uniform diameters.
1997	The first nanotechnology company is created, Zyvex.
1997	The first nanomechanical device based on the structure of DNA molecules was created by Ned Seeman.
2000	Researchers at Rice University have developed methods for transforming carbon nanotubes into rigid multicomponent structures.
2001	IBM researchers have developed methods for growing nanotubes.

Source: Medeiros and Mattoso, 2006

Chemistry plays a central role in the development and manipulation of nanomaterials. Since it works directly with atoms, molecules and particles, a very strong link between chemistry and Nanoscience and Nanotechnology (N&N) is to be expected. Although the focus of chemistry is very similar to N&N, they differ from each other. According to Fernandes and Figueiras (2008, p. 2205) "Chemistry deals with the structural composition and properties of matter at the atomic and molecular level. Nanosciences and nanotechnologies (N&N), on the other hand, encompass design, manipulation, production and assembly at the atomic and molecular level."

What this new technology does is give specific directions to reactions that take place on a nanometric scale and thus obtain favourable results. Thus, "nanotechnology is based on the development of appropriate techniques and tools to position atoms and molecules in previously established locations in order to obtain structures and materials of interest" (RIBOLDI, 2009, p. 04). Seen in this way, chemical knowledge serves to foster discussions about nanotechnology, since it is basically based on the manipulation of molecules and atoms.

In this way, issues related to the production of scientific knowledge should be addressed in lessons, as this clarifies the more practical side of chemistry, leading students to a contextual view of this science in the development of themes that are so significant for the future of humanity. The knowledge transmitted by the school must go beyond the classroom walls and reach society, as well as the necessary changes inherent in interpreting the scientific knowledge produced.

From this perspective, realising that a topic is important for society should awaken in the teacher a desire to discuss the positive and negative points with the students, who are the subjects of the teaching process. In the N&N approach, for example, it is extremely important to discuss with students improvements in health, such as the development of drug nanocarriers capable of interacting specifically with damaged cells, thus increasing patients' chances of being cured. According to Dime et al (2015, p. 1520), research into the development of "smart" drugs is turning its attention to "the treatment of cancer, inflammatory diseases, cardiovascular diseases, neurological diseases and the fight against the acquired immunodeficiency virus (AIDS)".

However, it is important to point out that the use of nanotechnological tools without proper follow-up and specific monitoring can cause damage at molecular levels to both human beings and the environment. If in industry materials have their properties modified, the by-products that can pollute the air, atmosphere and water have a high potential as polluters because they also have their properties modified. "This can be a great opportunity to work on this topic at school, through activities that enable students to become aware of the potential benefits and harms that a new technology can bring" (REBELLO et al, 2012, p. 04).

Today, little has been said about nanotechnology during secondary school lessons. On the other hand, society is constantly bombarded with news reports about this term (SILVA, VIANA and MOHALLE, 2009). Hence the concern to include topics that express the contributions of this emerging technology in chemistry syllabuses. However, it is through chemistry lessons that students can better understand the field of study of nanotechnology, given the interdisciplinary nature of the subject:

> [...] teachers, especially chemistry teachers, need to know that nanotechnology is not a new and separate field of knowledge, but involves concepts they have already mastered such as atoms and molecules, particle size, metric scales, specific surface area, adhesion, among others. What is really new is the way in which atoms and molecules are being manipulated to create new technologies, and this can be taught within their standards of knowledge (SILVA, VIANA and MOHALLEM, 2009, p. 178).

Teachers must bear in mind that, given their importance for students' education, technologies and their possible impacts must be understood during the development of scientific knowledge. In this way, chemistry's contribution to enlightening and establishing a critical sense in students is unquestionable. Through the intrinsic interaction that can take place in the classroom, students have the opportunity to ask questions, debate ideas and relate the knowledge acquired at school to the new technologies that surround them.

Referring to the National Curriculum Parameters for chemistry teaching, it can be seen that "the cognitive and affective competences and skills developed in chemistry teaching should enable students to make their own decisions in problem situations, [...]" (BRASIL, 2000, p. 32).

In chemistry, these cognitive skills of representation, communication, investigation, understanding and socio-historical contextualisation refer to the academic knowledge structures acquired in social interaction outside and inside school. Affective skills, on the other hand, relate to the interpersonal relationships inherent to human beings and important for the construction of knowledge and personal growth.

From this perspective, chemistry teaching should be able to develop competences and skills in students, since it is based on understanding the physical and chemical transformations that occur in the world around them, contributing to the construction of a world that is more articulated with social, cultural and environmental aspects.

In an increasingly globalised world, the emergence of technologies with the aim of improving life in society requires knowledge related to a wide range of political, economic, social and cultural aspects. This knowledge can be developed at school, and in the teaching of chemistry, the Curriculum Parameters direct it towards industrial production, the use of fertilisers, the production of new technologies, as well as the impacts arising from this knowledge. Thus, the Curriculum Parameters establish that:

> This knowledge requires, among other things, the skills and abilities to recognise the role of chemistry in the production system, to recognise the relationship between scientific and technological development and socio-political-economic aspects, such as the relationship between fertiliser production, agricultural productivity and environmental pollution, and to recognise the ethical and moral limits involved in the development of chemistry and technology, pointing out the importance of using environmentally clean industrial processes, pollution control and monitoring, and public disclosure of environmental quality indices (BRASIL, 2000, p.35).

We therefore realise that the conscious development of chemistry and its technologies must be discussed in the educational sphere with an emphasis on social and scientific ethics.

The school thus becomes a place to think about the direction of the new technologies that are emerging, including nanotechnology. According to Toma (2005, p.09) "the issue of nanotechnology cannot fail to be included in the education scenario, since the main players in this new era will be young students, who will need to be prepared to enter this labour market".

According to this current scenario, the chemistry teaching-learning process is directly related to the historical and social context in which the student is inserted, and so there is a lot of talk about contextualised teaching in which the content seen in the classroom must be related to the student's world. However, the criticisms that some authors in the field of didactics bring to the teaching of chemistry leave us wondering about the true objective of this discipline.

Despite the need to include new concepts and themes in chemistry teaching, transforming it into contextualised teaching in which the content seen in the classroom is related to the student's world, what we see is teaching that falls far short of this reflection.

According to Lôbo (2008), in chemistry teaching it is quite common for scientific results to be presented and passed on to students as ready-made scientific knowledge in the form of laws, theories and models. Thus, the author reflects on chemical-scientific knowledge in which its production processes are not emphasised. In other words, the understanding of content is restricted to the information that is passed on to students, as packages of knowledge that are closed to discussion and this "causes a series of undesirable developments in the pedagogy of this science, creating obstacles to its understanding" (LÔBO, 2008, p. 92).

From this perspective, teaching is tied to the mechanicism of memorising mathematical formulas and scientific concepts without any relation to the student's life. In other words, the ready-made knowledge that scientific research brings us does not stimulate the student's critical thinking.

How can students be encouraged to think about how science and technology develop equipment that manipulates matter on nanometre scales with extraordinary performance? The main answer to this question could lie in the process of teaching and learning through investigative research, in which the students themselves must mobilise themselves in the construction of their own knowledge, with the teacher as a mediator of this process.

According to the same author, "one of the aspects most discussed by science educators is the difficulty students have in understanding scientific concepts" (LÔBO, 2008, . 91). Contextualising chemistry with issues related to scientific and technological development encourages critical thinking skills in students. The concern on the part of some scholars is based on the model of society that schools want to form. If, on the one hand, students are insistently impressed by technological development, many of them fail to understand how this development favours or disfavours the construction of a better society.

In the specific case of nanotechnology, where modern society enjoys many of the resources it provides, such as ever smaller electronic devices with more functions, few people can understand how such tiny mechanisms can have such a practical impact on human life. These mechanisms, which work on a nanometre scale, stem from the possibility of manipulating atoms and molecules in a well-defined and specific way.

It is believed that it is in primary and secondary school that students can familiarise themselves with such abstract concepts and work with their teacher on the nanometric scale. In this way, students' cognitive structures are developed and they have the opportunity to see the world with a sharper and more critical eye (ELLWANGER et al, 2012).

Chemistry teaching can provide the means to develop students' perception that their everyday lives are related to the content and, due to the interdisciplinary nature of topics such as nanotechnology, they have the chance to develop in other areas of science such as physics and biology. "In this context, the teachers of the various disciplines also have an important role, which is to always guide and instigate the student to make all this interrelationship" (SILVA, VIANA and MOHALLEN, 2009, p. 172).

Contextualisation sets the student up to investigate like a researcher or scientist. In this context, where students themselves try to formulate their own hypotheses through guided research, the metaphor of the "novice scientist" is established. This is a model of learning as guided research, where the student formulates hypotheses and ideas about something, and tries to prove them in a way that is as similar as possible to the scientific concept, which science deals with in the formal moulds. However, it is essential that students are faced with a problem situation in order to encourage them to research and formulate their own ideas about the problem (NUNEZ et al, 2004). This subject will be returned to in the methodological strategies.

2.1.1 Scientific Ethics and Education: Implications, Applications and Perspectives of Nanotechnology

According to Ellwanger et al (2012), today's technology and science are definitely moving towards the nanometric scale, both through miniaturisation in electronics and through nanostructural assembly from atoms and molecules. In this way, education has a major contribution to make to the training of future researchers in nanotechnology, both in terms of scientific knowledge and the ethics of policies based on preserving the environment.

The school environment, as an academic trainer, must promote awareness that technological development must be based on policies of scientific ethics, because "the ethics of science is based on the construction of knowledge to produce benefits for man and the environment. In no way can science be treated in a commercial and profitable way, valuing only profits" (SILVA, VIANA E MOHALLEN, 2009, p. 177).

As a teaching practice, scientific ethics must be considered as something that is linked to teachers' pedagogical practices. Discussing the benefits of nanotechnology in human life cannot escape the ethical and moral precepts that govern scientific society. If on the one hand new technologies develop a country's financial capital, on the other they can cause irreparable damage to the environment. As research into nanotechnology is still in its infancy, its potential risks are still being studied, but it is thought that the tiny size of the particles increases their harmfulness dramatically, as they can interact more easily with neighbouring systems. Corroborating this thought, Paschoalino, Marcone and Jardim (2010) report the following:

> The first initiatives to gather information on the areas that need to be studied in order to assess the harmful effects of nanomaterials on health and the environment were carried out by the Royal Society and the Royal Academy of Engineers and by the American NSCT (a subcommittee of the National Science and Technology Council). In the UK, the Royal Society and the Royal Academy of Engineers published a study in 2004 on the environmental risks to health, safety and the ethical and social implications associated with the development of nanotechnology and the areas where regulation should be considered (PASCHOALINO, MARCONE and JARDIM, 2010, p. 427).

The technologies that emerge demand efforts both from the researchers who develop these new technologies and from government officials in a corporate sphere in which scientific ethics must be emphasised. Nanotechnology has been discovered and is still being researched further, but very little is known about its potential as a science and a technology. It is to be hoped that the development of nanomaterials structured for specific applications does not cross the ethical and scientific boundaries between Science-Technology-Society (STS).

Despite the potential properties of nanocomposites and nanomaterials for improving quality of life, researchers do not fully understand the potential risks they can cause to the environment and to life in society. For this reason, "as nanotechnology gains visibility and presence in society and the media, questions about its social, environmental and ethical impacts will begin to gain more and more visibility among the lay public" (SILVA, 2004, p.06).

For Schulz (2009), the most serious of the implications of this new technology for the environment is the dilemma of nanotoxicology, which in general terms can be understood as the in-depth study of the action of nanoparticles developed for industrial applications on living organisms. He also points out that between 2006 and 2008, the production of nanocompounds tripled from 200 to 600, forcing a group of toxicologists to spread around the world and create high-speed analysis protocols with the aim of testing the toxicity of hundreds and thousands of nanotechnological materials simultaneously.

According to Silva (2003), with the sudden emergence of nanotechnology, society is forced to adapt and integrate with a new model of scientific technology in a short space of time like never before, and this should raise concerns in the field of ethics, since the problematisation of this discussion has direct implications for the direction society should take.

This situation has an impact on people's lives and creates a modern society that is subjected to waiting for technological mechanisms to provide quality of life. People need more up-to-date electronic devices with extended memory and longer battery life, and they see new technologies as more agile in the field of information technology, for example.

With technological development, "we don't know for sure to what extent there is concern about the compatibility between the use of new means of technical control of man and society, and the respect that is due to the human person" (SILVA, 2003, p. 10). This reaffirms the importance of educational policies that support the idea that nanotechnological development needs to be based on ethical principles that guide society. Development brings wealth and growth to the country, but it doesn't necessarily require human ethical principles to be sacrificed.

Questions about scientific ethics and morality should therefore be gradually introduced and discussed in the classroom, highlighting the causes of concern about the direction of scientific research in the country. Reinvigorating this thought, Siqueira-batista et al (2009, p.83) report that:

> In fact, the questions surrounding the production of nanomaterials, their application to health and the environmental impact of nanotechnology have as their backdrop the consequences - good or bad, splendid or catastrophic - on the planet. This is the aspect that can be debated by science students, interestingly combining scientific concepts - worked on interdisciplinarily due to the very nature of nanotechnology - with the argumentation required by the human sciences, in terms of ethics and politics (SIQUEIRA-BATISTA et al, 2009, p.83).

According to the Brazilian Agency for Industrial Development (ABDI, 2008), sectors related to nanotechnology or branches linked to it, such as nanobiotechnology, must outline strategies and draw up specific laws and policies to support the production and commercialisation of technologies developed taking into account market, ethical and social aspects.

Although it is not very developed in Brazil, nanotechnology is being worked on in industries and in centres that promote technological research. According to Fernandes and Filgueiras (2008, p.2212) the challenge of N&N in Brazil shows "scientific production in the area and innovation indicators that, although not as good, already show potential for growth, and it is essential that all advances are accompanied by measures to publicise science and social participation". From this perspective, the Brazilian government is carrying out research in technological areas, exploring raw materials used in the manufacture of nanocomposites, such as nanoporous materials like clays (BORSCHIVER, et al 2005).

The applications of nanomaterials are part of studies and research at the current juncture of rising nanotechnological development. The main interest in developing research in the field of nanotechnology lies above all in the advantages they bring to society. Thus "[...] the fundamental chemical and physical properties of materials depend on their size, or, in a looser language, that they maintain a 'complicity' with it, a complicity that is the key to all nanoscience" (ALVES, 2004, p.27).

With science's interest in the nanometric properties of compounds, research into carbon on a nanotechnological scale has developed. It is known that the properties of compounds are modified when working on a nanometre scale. This potentially increases their use and application in various areas of scientific knowledge, such as "catalysis, biological sensors, optoelectronic materials, controlled drug release and bioencapsulation". (SCHUCH, BEVILAQUA and FAGAN, 2007, p.96). Table 02 below summarises the main applications of nanotechnology.

Table 02 - Main applications of nanotechnology

ENERGY PRODUCTION	Photovoltaic and photoelectrochemical cells, organic light-emitting devices (OLEDs), applications of carbon nanotubes in solar cells, nanocatalysts for hydrogen production, fuel cells (TOMA, 2005).
PRODUCTION AGRICULTURE	Development of nanoporous zeolites for the slow and controlled release of water, fertilisers and agro-defensives; herbicide-releasing nanocapsules; nanosensors for monitoring soil and plant development; and magnetic nanoparticles to combat soil contamination (TOMA, 2005).
ENVIRONMENT	Selective membranes that can remove contaminants or even eliminate salt from water; nanostructured devices capable of removing pollutants from industrial effluents; characterisation of the effects of nanostructures on the environment; significant reduction in the use of materials and energy; reduction of pollution sources (ALVES, 2004).
PROGRESS IN MEDICINE	New drugs based on nanostructures, drug delivery systems that target specific points in the human body. (ALVES, 2004). "lab-on-achip"; "arrays" of nanosensors, magnetic nanoparticles and "quantum-dots" for clinical diagnosis; dendrimer-antibody complexes for HIV-1 and cancer diagnosis (TOMA, 2005; TEDESCO, 2006).
INDUSTRIES ELECTRONICS	Data recording using nanolayers and quantum *dots;* flat screens; wireless technologies; new devices and processes within all aspects of information and communication technologies; increased data processing speeds and storage capacities that are at the same time less expensive than the current ones (ALVES, 2004; SCHULZ, 2009).
MONITORING HEALTH	Sensors for glucose, CO2, cholesterol; sensors for in situ homeostatic monitoring (TOMA, 2005).
EXPROLARATION SPACE	Building lighter space vehicles (ALVES, 2004).

DEFENCE	Detectors and remediators of chemical and biological agents; increasingly efficient electronic circuits; much more resistant nanostructured materials and coatings; lighter fabrics with self-repairing properties; new blood substitutes; miniaturised security systems (ALVES, 2004).

2.2 CHEMISTRY TEACHING IN BRAZIL IN ITS HISTORICAL CONTEXT

In this section, chemistry teaching will be addressed in its historical context to support the current situation in Brazil, both in basic education and in higher education. To do this, it is necessary to give a historical overview of how this subject emerged in the country, the first higher education courses and the prospects for the development of research in teaching.

The first records of the history of secondary chemistry teaching (now primary education) date back to 1862, when it was taught alongside physics. It wasn't until Rocha Vaz's educational reform in 1925 that the two subjects were taught separately, although chemistry classes were only taught in the last two years of secondary school. Perhaps the reason for this lies in the educational legacy left by the Jesuits, which was characterised by a scholastic, literary model and a lack of interest in scientific studies. Even with the expulsion of the Jesuit priests from Brazil in 1759, science teaching was neglected and a literary and humanistic spirit prevailed in schools to the detriment of scientific subjects (SCHNETZLER, 2010).

The first degree course in chemistry in Brazil dates back to 1918, offered by the Institute of Chemistry in Rio de Janeiro, and in 1920 the degree course in Agricultural Industrial Chemistry was created at the School of Agriculture and Veterinary Medicine (OLIVEIRA and CARVALHO, 2006).

After the various educational reforms that Brazil has undergone, we realise that..:

> Since the Francisco Campos reform (1931-1941), basically the same objectives have been proposed for secondary chemistry education, namely: to promote the learning of the general principles of chemical science; to emphasise its experimental nature and its relationship with students' everyday lives, purposes which, in my opinion, sought and seek to give meaning to the compulsory nature of that teaching, whether in secondary school (19311941), in the scientific course (1942-1960), in secondary school (1961-1995) and currently in secondary school (SCHNETZLER, 2010, p. 56).

Despite some obstacles, the teaching of chemistry has made great strides in research and investigation with the creation of various organisations to promote scientific research in Brazil. Some of these initial milestones are: the creation, in 1951, of the National Research Council (CNPq) and the National Coordination for the Improvement of Higher Education Personnel (CAPES); the creation, in 1977, of the Brazilian Chemical Society (SBQ); in 1981, the creation of the National Chemistry Support Programme (PRONAQ) which later resulted in PADCT, the main Brazilian Chemistry Support Programme (ANDRADE et all, 2005).

Schnetzler (2010) points out that in addition to the creation of these bodies, research into chemistry education has been favoured by the National and Regional Meetings on Chemistry Teaching (ENEQ). The first ENEQ was held in 1982 at Unicamp and since then the meetings have only tended to increase in number of participants and in the knowledge of teachers and undergraduates.

Another important milestone was the creation of the journal Química Nova na Escola (QNEsc) in 1995, aimed at chemistry teachers and undergraduates. In recent years, its range of publications has included the names of researchers who have contributed significantly to the dissemination of research in Chemistry Education in Brazil, such as Attico Chassot, Letícia Parente, Luiz Otávio Amaral, Luiz Roberto Pitombo, Mansur Lutfi, Maria Eunice Ribeiro Marcondes, Otavio Maldaner, Roberto Ribeiro da Silva, Romeu Rocha-Filho and Roque Moraes, among others.

QNEsc magazine currently publishes unpublished scientific articles every quarter in the following sections: Current Events in Chemistry, Outstanding Scientific Concepts, Chemistry and Society, History of Chemistry, Classroom Reports, Experimentation in Teaching Chemistry, The Student in Focus, Research in Teaching, Chemical Element, Chemical Education and Multimedia and Open Space.

In addition to QNEsc, the SBQ also includes Química Nova and the Journal of the Brazilian Chemical Society. However, it is QNEsc that provides a wide-open space for educators to discuss and reflect on the teaching and learning of chemistry.

An analysis of these documents reveals that one of the concerns of educators and researchers is the heterogeneous range of higher education offered by higher education institutions. The concern is about the quality of the professionals being trained to work in the labour market, especially in the Licentiate and Bachelor's degree courses. Among these reports, we can cite Zucco (2005, p. 13) who pointed out:

> [...] the courses aimed at training chemistry teachers do not have the right conditions, and are therefore co-responsible for the poor quality of their teaching.
> quality of high school graduates. Another conclusion is that the courses run by federal HEIs and some state HEIs are the best. This is certainly due to the qualifications of the teaching staff on these courses and the involvement of their lecturers in scientific activities. [...] changes need to be made to the didactic-pedagogical projects of the courses so that they offer a solid education in chemistry, but one that is comprehensive and generalist enough for the chemist to be able to develop in more than one direction. There are many challenges facing us in this task of training chemists: i) implementing the Guidelines; ii) improving the teaching qualifications of university and high school teachers; iii) redefining professional training in order to train graduates who are familiar with new technologies and have an entrepreneurial spirit; iv) training postgraduates who can work in the industrial sector; v) seeking to transform Brazil's basic chemical industry into a specialised chemical industry; vi) seeking to change professional attributions (in order to eliminate the verticalisation of attributions) and the recognition of postgraduation as a professional qualification.

With all the investment in research in chemical education, whether in professional training and qualification or in the creation of means of conveying the knowledge produced, Schnetzler (2002, p.22) reveals that "the contributions of research to improving the teaching-learning process still don't reach the majority of teachers who actually make teaching happen in the schools of this immense country". According to the author, one of the major factors highlighted by researchers in the field that contributes to this is the lack of encouragement for educational research during initial and continuing training.

In this context, the search for teacher qualification and professionalisation is currently the focus of several research studies in postgraduate programmes in Brazil. At the centre of this research are educators in training or already working in the last stage of primary and secondary education. The prospects are that they will be able to learn about existing educational research in order to intervene in teaching and learning difficulties in the classroom.

2.2.1 Training Chemistry Teachers

Some research into the initial and continuing training of chemistry teachers reveals that after several years the courses that train these professionals have lost credibility and in some extreme cases have even closed down. According to Vianna, Aydos and Salgado (1997) there are a number of factors that add up to highlight this reality, such as:

> The lack of integration between specific content and education subjects, the fragmentation of content and discrimination against teachers and students by colleagues from Bachelor's and related courses, the low level of knowledge of students entering the course, the lack of professional prospects due to low pay and poor working conditions in secondary schools, combined with the disregard for education shown by our governments (VIANNA, AYDOS and SALGADO, 1997, 213).

Notwithstanding this reality, theorists in the field of teaching and education share the view that there will only be real improvements in the quality of science teaching when there is investment in the initial and ongoing training of teachers who teach grades 5 to 8 of primary and secondary school, or more recently grades 4 to 9 of primary and secondary school. These investments range from incentives to attract professionals to education, such as improvements in school infrastructure and better salaries, to solid training at universities or in specific teacher training centres, as is the case in some countries (MALDANER and SANTOS, 2010; CARVALHO and GIL-PÉREZ, 2011; SILVA and OLIVEIRA, 2010).

Based on these premises, an explanation can be made establishing some parameters for qualitative analysis of the initial and continuing training of chemistry teachers. One of these parameters is the objective of chemistry teacher training courses, which according to Silva and Oliveira (2010, p.43) would be:

> The aim of chemistry degree programmes is to train teachers to work in basic education. This training must include numerous aspects inherent to the training of a good teacher, such as knowledge of the content to be taught, curricular knowledge, pedagogical knowledge about the school subject of chemistry, knowledge about the construction of the

scientific knowledge, specificities about the teaching and learning of chemical science, among others (SILVA and OLIVEIRA, 2010, p.43).

It is thus argued that chemistry teachers should have solid knowledge of the content to be taught, but should also have pedagogical knowledge and knowledge of the specifics of the teaching and learning processes in this subject. This critical reflection on teacher training makes it possible, according to Carvalho and Gil-Pérez (2011, p. 73), "to break with the erroneous but widespread idea that teaching a subject is a simple job, for which it is enough to have a higher level of knowledge than the students".

In addition, it is established that chemistry teachers must have knowledge about the construction of scientific knowledge, how it has been established over the years in society and what its contributions and implications are. All these reflections refer to the thoughts of Bachelard, a theorist of the epistemology of science who influenced and still influences many works and research in science education. According to Bachelard (1996), the search for the construction of scientific knowledge starts with the formation of the scientific spirit, and when it comes to education, especially science teachers, the epistemologist reflects on the epistemological and pedagogical obstacles to disclosure:

> I find it surprising that science teachers, more than others if that were possible, don't understand that someone doesn't understand. There are few who have delved into the psychology of error, ignorance and thoughtlessness. [...]. Science teachers imagine that the spirit begins like a lesson, that it is always possible to rebuild a flawed culture by repeating the lesson, that a demonstration can be made understandable by repeating it point by point. They don't take into account the fact that adolescents come to physics lessons with empirical knowledge that has already been built up: it's therefore not a question of acquiring an experimental culture, but of changing an experimental culture, of breaking down the obstacles that have already been sedimented by everyday life (BACHELARD, 1996, p. 23).

From this perspective, the epistemological nature of scientific knowledge is consolidated in science teacher training courses. It is the epistemology of science that makes it necessary to reflect on the veracity of scientific facts and theories about a given phenomenon from the perspective that man produces scientific knowledge and this cannot be neglected. In the context of teaching, this thinking is evident when Bachelard states that the student

comes to physics class with empirical knowledge already built up and the teacher's job is to change some misconceptions, and thus change their experimental culture.

Maldaner and Santos (2010) state that Bachelard's conceptions of the production and validation of science are based on the concepts of real given and real produced. Thus, there are two production processes through which scientific culture is formed: by the student's everyday life (real given) and by the rupture of the first perceptions and with the student's everyday life, in other words, the scientific real (real produced). In chemistry teaching, this rupture between the real given and the real produced is evident in the following contents:

> [...] they are the atomic-molecular structures, with which we think of the material world, the chemical equation, with which we understand and control the transformations of materials, and chemical representations of all kinds, which give the basic idea of the composition and properties of materials (MALDANER and SANTOS, 2010, p. 341).

In this way, the authors highlight part of the content that exemplifies and explains Bachelard's epistemological thoughts, and in this context, the importance of chemistry in understanding the world and the student's daily life can be seen. In other words, it can be said that, in a simplistic way, Bachelard's epistemological considerations can be achieved when the teacher is able to critically reflect on how chemistry, with its own languages, contributes to explaining observed phenomena.

However, teacher training is not limited to undergraduate courses and in this context, Silva and Oliveira (2010, p.45) emphasise that "we understand that teacher training is a continuous process that does not begin, let alone end, in an undergraduate course, such as a degree". Ongoing training ranges from courses to improve teaching practices, through lato sensu specialisation courses to stricto sensu master's and doctorate courses. In addition to this knowledge, teacher training also includes teaching knowledge, which goes beyond an academic approach and involves the personal, professional and organisational dimensions of the teaching profession (ALMEIDA and BIAJONE, 2007).

Another very important parameter to be analysed in chemistry degree courses is the specific training offered by Higher Education Institutions (HEIs). The teacher's mastery of the content to be taught is, by far, seen by educators in the field of chemistry.
science teaching, as the basis that guides the teacher's pedagogical actions. Neglecting specific training during the academic life of an aspiring teacher is like putting any professional from another area of knowledge to teach a subject they haven't mastered.

However, Furió and Gil-Pérez (1989) point out that in some teacher training courses there are two erroneous trends that are far apart: one focused on the transmission of scientific knowledge (usually at the beginning of the course) and the other emphasising general psycho-pedagogical preparation (usually at the end of the course). In other words, there is no integration between the content-orientated proposals and the didactic proposals. This gap between specific training and didactic-pedagogical training jeopardises the preparation of professional teachers.

Carvalho and Gil-Pérez (2011, p. 23) set out in table 03 that in addition to knowing the subject content, other professional knowledge required for teacher training goes beyond what is usually covered in university courses at HEIs.

Chart 03: Scope of scientific knowledge of the subject to be taught

A. Knowing the problems that led to *the* construction of scientific knowledge (without which this

knowledge appears as arbitrary constructions). Knowing, in particular, what the epistemological difficulties and obstacles were (which is an essential aid to understanding the students' difficulties).
B. Know the methodological guidelines used in the construction of knowledge, i.e. the way scientists approach problems, the most notable characteristics of their activity, the criteria for validating and accepting scientific theories.
C. Know the interactions between Science/Technology/Society associated with this construction, without ignoring the generally dramatic nature of the role of the social role of Science; the need for decision-making.
D. Have some knowledge of recent scientific developments and their prospects, in order to be able to convey a dynamic, non-closed view of science. Also acquire knowledge of other related subjects in order to be able to address related problems, the interactions between different fields and the processes of unification.
E. Knowing how to select appropriate content that gives a correct view of science and that is accessible to students and likely to interest them.
F. Be prepared to deepen your knowledge and acquire new ones.

Source: Carvalho and Gil-Pérez, 2011

Given what the authors have said, there is an understanding of the scope of content knowledge that goes beyond what is generally presented in the specific curricula of chemistry degree courses. What Carvalho and Gil-Pérez (2011) propose is a critical-reflective stance during training from the perspective of turning to the reality of the classroom, such as understanding what epistemological obstacles students are likely to have in relation to a particular piece of content or subject studied.

Libâneo (2011, p. 84) corroborates this thought when he states that "the idea is that the teacher can 'think' about their practice, or in other words, that the teacher develops the capacity to reflect on their own practice".

For a comparative analysis, Silva and Oliveira (2010, p.49) show (table 04) the common curricular structure, prior to 2006, of the first four semesters of a degree course in Chemistry at a university in São Paulo. It can be seen that there is no prospect of critical-reflective pedagogical training on knowledge of the science of chemistry due to the rigid and closed structure of the subjects in these first four semesters. Only two subjects favour didactic-pedagogical discussion of science, but in a general rather than specific way.

Table 04: Typical Curricular Structure of a Degree in Chemistry

CURRICULUM STRUCTURE	
1ST YEAR	
1st Semester	**2º Semester**
Differential and Integral Calculus I (04)	Differential and Integral Calculus II (06)
Introduction to Computer Science (04)	General Physics I (04)
Introduction to the Study of Chemistry (04)	Experimental General Chemistry II (04)
Experimental General Chemistry I (04)	General Chemistry II (04)
2º YEAR	
3º Semester	**4º Semester**
Foundations of Education (04)	Elements of Geology and Mineralogy (04)
General Physics II (04)	General Physics III (04)
Qualitative Analytical Chemistry (08)	Quantitative Analytical Chemistry (08)
Basic Inorganic Chemistry (04)	*History and Philosophy of Science (04)*

Source: Silva and Oliveira, 2010

This case study, however, does not reveal the reality of all HEIs in Brazil, but rather that initial and continuing teacher training courses have become more concerned with pedagogical training. One example is the IFRN, whose chemistry degree programme was changed in 2012, giving greater coverage to subjects in the Didactic-Pedagogical and Epistemological core areas. It is also worth highlighting the institution's concern with academic practice by proposing integrative projects in its curriculum, which give its students earlier contact with academic research. From now on, the fifth term sees no specific core subjects, only didactic and pedagogical subjects, weakening the integration between the specific and pedagogical cores.

The last parameter taken into consideration here by the proposed qualitative analysis concerns the pedagogical training content in chemistry degree courses, which emphasises the importance of the didactics of this science in the teaching and learning process. The great emphasis given to specific disciplinary content can jeopardise the quality of teaching because a good teacher is not one who only masters the content, but one who knows the subject and also knows how to teach it to their students.

Pedagogical training will therefore enable teachers to "formulate and present content in a way that makes it comprehensible to students, including analogies, illustrations, examples, explanations and demonstrations" (ALMEIDA and BIAJONE, 2007, p.288). In chemistry, considered by many students to be a difficult subject to understand because it deals with mostly abstract phenomena, teaching strategies play a central role in the teaching and learning process.

However, what happens in chemistry degree courses is a disconnect between specific and pedagogical content, in other words, "there seems to be no common objective among the teachers who teach the subjects, and this aspect seems to be a consensus in several degree courses and one of the aspects that most favours the poor training of chemistry teachers" (SILVA and OLIVEIRA, 2010, p. 55).

Initial teacher training should be articulated as practices in critical-reflective activities and this articulation needs to happen at the beginning of the course and not just in the regency internship subjects. However, we realise that

> Various studies have pointed out the problems of initial and continuing teacher training. They criticise the rigidity of the curriculum and methodology of training courses and the disconnection from practice. Continuing education initiatives, usually in the form of "training", have been widely criticised, as some studies have shown. What's more, teachers face disparaging criticism from various quarters, leading to an uncomfortable discrediting of their profession. However, it is true that quality general training for students depends on quality training for teachers (LIBÂNEO, 2011, p. 82-83).

It is therefore believed that quality teaching presupposes the provision of good training in degree courses, in which their curricula establish a balanced articulation between the subjects of specific knowledge and the subjects of pedagogical knowledge, from the perspective of critical-reflective training in pedagogical practices. In addition, teachers' professionalisation should be encouraged and valued through policies that allow educators to seek continuing training to improve their practices.

2.3 METHODOLOGICAL STRATEGIES USED IN TEACHING NANOTECHNOLOGY

There is a consensus among teaching researchers that science and the new technologies produced today must be made accessible to primary school students. Thus, didactic-methodological strategies and tools are being developed with the aim of stimulating the learning of science content by bringing the production of new technologies into classroom discussions (AFONSO, 2011; TOMA, 2005; SCHULZ, 2009).

One example of teaching strategies that is gaining prominence among science education researchers is investigation as an activity and argumentation as a way of discussing and understanding the products of the science being investigated (CARVALHO, 2013; BELLUCO and CARVALHO, 2014; ZÔMPERO and LABURÚ, 2011; NUNEZ and RIBEIRO, 2004). Nanotechnology is revolutionising the way we see and understand the limits of science in contemporary society (SCHULZ, 2009; TOMA, 2005; ALFONSO, 2011).

Relating new technologies such as nanotechnology to science education has been a constant concern among the world's most developed countries (SANNOMIYA, HONÓRIO and PEREIRA, 2010). According to the Brazilian Agency for Industrial Development (ABDI), the growth of nanotechnology is taking on ever greater proportions in world-leading countries such as the United States of America, Japan and the countries that make up the European community, investing in initiatives to provide competitiveness among companies in the technological field.

This incentive is based on the improvements that nanotechnology can offer society, such as: agriculture, energy, quality of life, environmental preservation, public health and many others. Considering these perspectives, several countries are concerned about the education of children and young people, investing in science education aimed at teaching nanotechnology, such as Thailand, England, Germany, the United Kingdom and others (SANNOMIYA, HONÓRIO and PEREIRA, 2010).

In Brazil, some initiatives have already been promoted with the aim of publicising and providing the public with experiences, albeit timid, in this field of scientific and technological development. *NnooAeentraa"* is one such initiative. It is a science fair organised by the exploratory science museum of the State University of Campinas (Unicamp) together with students and researchers (ALFONSO, 2011). A

The aim is to teach students and visitors some of the concepts and applications of nanotechnology through virtual educational games.

Proposals are needed to make it possible to include nanotechnology in the formal teaching of public schools, with the aim of improving the quality of teaching and learning of scientific and technological concepts. In this educational scenario, it is important to study nanotechnology in primary and secondary schools, as young people are interested and motivated to study new technologies and so can prepare themselves to enter research and even the labour market (TOMA, 2005; ALFONSO, 2011).

With this in mind, a lot of research and literature proposes working on nanotechnology alongside the chemistry, physics and biology subjects offered in basic education. Chemistry content covers subjects such as atoms, molecules and particles that are directly related to nanotechnology.

This new technology can facilitate and contribute to teaching that is more attractive, stimulating and capable of training students as citizens who are able to reflect on the new technologies with positions that are favourable or against them, but with a coherently grounded position. The initial training of secondary school teachers is therefore essential (LIMA and ALMEIDA, 2012).

Therefore, understanding that a topic is important for society should awaken in the teacher a desire to discuss the positive and negative points with the students, who are the subjects of the teaching process. According to Silva, Viana and Mohallem (2009), nanotechnology and nanosciences have been covered very little during secondary school lessons. On the other hand, the authors state that society is constantly bombarded with news reports about these terms in the media.

This section discusses three methodological approaches to teaching nanotechnology, with an explanation and didactic analysis to justify their choices. The first approach is characterised by investigative activities that associate the teaching of nanotechnology in such a way as to encourage investigation in the search for a solution to a problem.

In the second, teaching is proposed with the help of illustrative teaching materials, such as nano-art (screens of nanometre-scale materials obtained by scanning electron microscopy); video and documentary proposals that dynamically show basic concepts of nanotechnology and a comic book that gives students greater interactivity with the illustrations and the scientific concepts covered.

Finally, the proposal to work on the construction of concept maps based on the theme is presented, both to introduce concepts and to revise them before the start of the lesson. The proposal is also demonstrated as a way of assessing the students' knowledge.

2.3.1 Teaching science through investigative activities

Inquiry-based science teaching emerged at the beginning of the 20th century on the North American educational scene, with thinker and educator John Dewey as the main influencer of this teaching proposal. Its main objective was to bring scientific knowledge produced in research centres closer to the reality of the classroom through the scientific method (ZÔMPERO and LABURÚ, 2011).

However, Trópia (2011) points out that the perspective of inquiry-based teaching proposed by Dewey in the USA, which began to influence curriculum reforms in Brazil in the 1950s and 1960s, had a conception of science as neutral and detached from society. It wasn't until the end of the 20th century that researchers in the field of science teaching woke up to this gap in thinking and in written teaching materials, based on the importance of the relationship between science and society. In this way, teaching would no longer be recognised for its passivity, but rather for the active production of knowledge by the subjects that make up the teaching and learning process.

Following Dewey's thinking and ideas, several researchers have come to the conclusion that enquiry teaching has a number of stages that must be followed in order to achieve the main objective, which is to give students moments as researchers.

In chemistry teaching, for example, investigation has taken on relevant scope in experimental activities that can have this investigative bias in laboratories as well as in classrooms. Associating nanotechnology with investigative experimental activities becomes necessary because it puts the student at the centre of the teaching and learning process, developing skills and relevant scientific concepts, such as: intermolecular forces acting at atomic levels, new properties of matter, the evolution of modern science, among others.

When it comes to teaching chemistry, investigative activities can be carried out in the classroom simply by reading a scientific text, or even with the help of a teaching video that provides the expected result planned by the teacher. However, it is necessary to recognise the importance of such an activity for the effectiveness of the results. For example, the text or video being worked on should provoke some kind of concern in the students that encourages them to research or investigate (this is usually achieved when there is an exciting problem to be solved).

Furthermore, it is essential that the investigative activity correlates with concepts and content specific to chemistry, uses scientific language appropriate to the level of the students and manages to integrate scientific knowledge with society, so that the teacher can qualitatively measure degrees of openness or difficulty in the activities that accompany the development of skills and ability to solve the problems presented. This strategy of monitoring performance during activities can be an effective way of assessing student performance in a non-traditional way.

However, the teacher needs to realise that "progress in student performance, autonomy and other skills developed through investigative activities are not immediate" (FERREIRA, HARTWIG and OLIVEIRA, p. 104, 2009). In other words, it is a methodological approach in which qualitative assessment (active participation, research and problem-solving skills, group work, the use of scientific methods in research, etc.) takes precedence over quantitative assessment, which prioritises grades regardless of the means by which students achieve them.

With regard to the teaching of nanotechnology, there is great potential in chemistry classes through experimentation with an investigative bias, because although it involves concepts that refer to modern technologies and are far removed from the reality of the classroom, it is possible to work in an alternative and contextual way.

In this context, Tasca et al. (2015) highlight the development of nanotechnology skills and concepts in secondary education as a way of showing that the development of new technologies has driven the search for new materials with previously controlled properties. To this end, they make use of experimentation in which the students actively lead all the steps of the experimental activity to synthesise superparamagnetic nanoparticles (NPMag), as well as a practical application of them. The applicability of NPMag has aroused the interest of researchers all over the world, because "due to quantum size effects and large surface area, they have some of their magnetic properties drastically altered, exhibiting quantum magnetisation tunneling phenomena and superparamagnetic behaviour" (TASCA, et al, p. 236, 2015).

As an investigative teaching activity, the synthesised NPMag were modified by giving them an organic layer of oleic acid capable of interacting with organic compounds present in oil derivatives. Considering the major oil rig accidents that history has told us about, with oil spills and consequent environmental disasters, the teacher can contextualise and encourage students to research/investigate the application of the synthesised nanoparticles for removing spilled oil.

2.3.2 Illustrative teaching materials

The teaching of nanotechnology in chemistry classes can be mediated by illustrative teaching materials that show nanomaterials in a different way to the forms that ordinary materials usually take. In their quest to unravel the mysteries behind the scientific nature of nanomaterials, scientists and researchers have discovered a fascinating world with varied shapes and colours that can offer beautiful visual spectacles in nano-art screen exhibitions.

This project emerged from images obtained by high-resolution electron microscopes by researchers from the Multidisciplinary Centre for the Development of Ceramic Materials (CMDMC) and the National Institute for Materials Sciences in Nanotechnology (INCTMN), who combined scientific knowledge and art. The project, which has been running since 2008 in partnership with USP, São Paulo State University (Unesp), the Federal University of São Carlos (UFScar) and the Institute for Energy and Nuclear Research (IPEN - a state organisation associated with USP), has already been presented in the United States, Israel and at meetings of the Brazilian Society for the Advancement of Science (NAOE, 2015).

According to Naoe (2015), the CMDMC project also reaches out to students in public schools to give them access to science and culture through short courses offered to secondary and primary school students, while teachers are usually given lectures and mini-courses on the Evolution of Technology and Energy Demands with their respective impacts on the environment. This idea promotes scientific dissemination about nanotechnology and helps scientific education, which according to Schulz (2008) is still incipient in society as a whole.

Some images are shown below to illustrate the art hidden behind nanotechnology. The images are part of some exhibitions from the "Nanoart and science as art" projects that have already been seen or accessed on the internet by thousands of people around the world. As a methodological strategy for teaching nanotechnology in chemistry lessons, the teacher can ask the students to set up exhibitions in the school about the nanoworld of materials. In addition to the interaction promoted by teamwork, knowledge about nanotechnology will emerge as a result of the research carried out by the students when they set up their "displays". What's incredible about this work is that the images mess with visitors' thinking when they try to compare them with something they already know.

Despite the close resemblance to the image of carnations, figure 02 shows zinc oxide nanostructures obtained by scanning electron microscopy and then coloured using image editor programs. This image is part of the **collection of the 2010 "SCIENCE AS ART" competition organised by the** Materials Research Society (POLETTO, 2012).

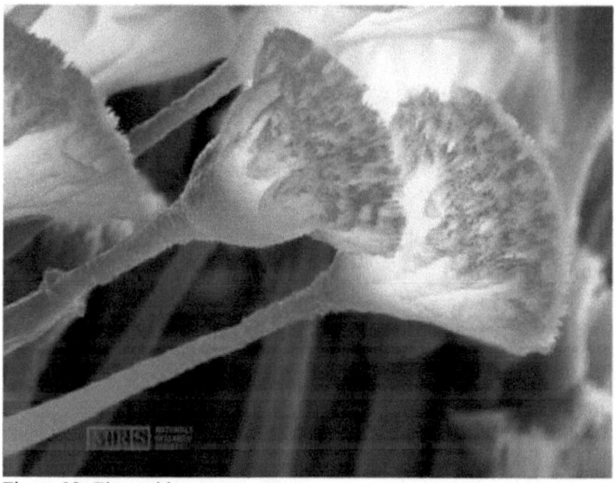

Figure 02: Zinc oxide nanostructures
Source: Poletto, 2012

Figure 03, although it looks like a pink ball surrounded by long, thin fingers with painted nails, is actually a polystyrene microsphere (diameter = 2.5 micrometres) surrounded by epoxy resin fibres (diameter = 250 nanometres), obtained by scanning electron microscopy and then coloured with image editor programs. This image is part of the collection of the **2008 "SCIENCE AS ART" competition** organised by the Materials Research Society (POLETTO, 2012).

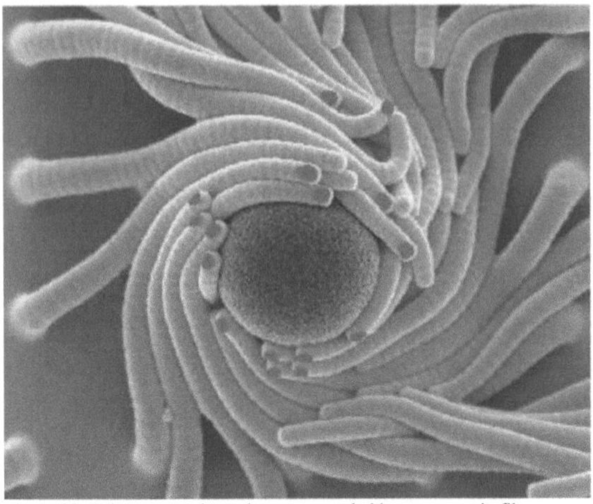

Figure 03: Polystyrene microsphere surrounded by epoxy resin fibres.
Source: Poletto, 2012

When you look at figure 04, although it looks a lot like coloured balls of wool, it is, in the nanometric universe, a representation of copper oxide obtained by scanning electron microscopy and then coloured with image editor programs. This image belongs to the exhibition collection of the Nanoart project (NAOE, 2015).

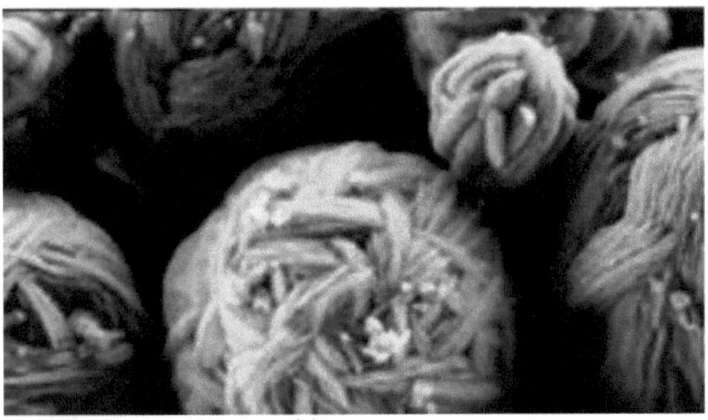

Figure 04: Copper oxide nanoparticles
Source: Naoe, 2015

Another way teachers can teach nanotechnology is through videos and documentaries on the subject. It should be emphasised that the activity must not be detached from the objectives set out in the school plan, i.e. it must not be seen as a way of occupying the space of one or two lessons, but must be based on the enrichment of the information covered by the documentary. The activity of watching a video or documentary is seen as didactic if, in some way, the students perform satisfactorily. This learning can be observed through debates and discussions on the topic covered in the video-documentary, in which the students will show whether or not they have mastered the subject; the creation of a jury that favours different points of view from the students, with space for them to present their ideas and try to convince others; and even summaries, fiches and notes can help the teacher in the task of evaluating their teaching practices.

Bassoto (2011) suggests the Discovery Channel science documentary on nanotechnology, entitled "Fantastic Journey - Through the Human Body in Search of a Cure", which lasts 50 minutes. But there are other documentaries on nanotechnology, such as the edition of the programme "Matéria de capa - Nanotecnologia", a documentary shown on TV Cultura lasting 29 min.

In addition to the above, many science fiction films use fantastical and fictitious aspects of nanotechnology in their trilogies. The interesting thing is the possibility of mixing fiction with reality and comparing the development of new technologies from a few decades ago with the present day. In this way, students will realise that much of what was considered fiction before is now outdated technology.

Finally, the teaching of nanotechnology in chemistry classes can be mediated by comic strips in which the structural and dynamic aspects of nanotechnology are mixed with the creativity and imagination of children and adolescents, favouring the teaching and learning activity.

However, for a long time comic books were stigmatised by the "thinking" social class, who saw them as the cause of children's and adolescents' educational problems, the dulling of logical thinking and even damaging to their readers' social and emotional relationships, as they provided them with a plunge into an imaginary world of abstract and fanciful ideas. But in the European cultural environment, there began to be an awakening to the rediscovery of comics and the unfounded preconceptions of parents and educators were being overturned, giving way to thinking of comics in their own positive light, as happened with detective literature and science fiction (VERGUEIRA et al, 2014).

The proposal to work on nanotechnology through comic books brings dynamism to teaching activities and provides students with a universe of knowledge that is different from that already known in textbooks. These differentiated resources break up the often tedious school routine by providing discovery mechanisms in instruments (comic books) that most students have already had access to. In this way, as well as arousing the student's interest, comics are important from a sociological point of view, as they are a means of playful communication and release tensions that affect a large proportion of children and adolescents all over the world (REIS, 2001).

With this in mind, a comic book was selected to help chemistry teachers teach and learn about nanotechnology and was analysed in terms of the scientific concepts presented, the complexity of the information and the chemistry content that can be covered in lessons. The comic book analysed and brought as an example for teachers to use is available free of charge electronically. It is the collection Nanotecnologia em quadrinhos n. 2 (2010). It is the result of the project "Impacts of nanotechnology on workers' health and the environment", which began in 2007 at FUNDACENTRO in partnership with RENANOSOMA (Research Network on Nanotechnology, Society and the

Environment), DIEESE (Inter-Union Department of Statistics and Socioeconomic Studies), DIESAT (Inter-Union Department of Studies and Research on Health and Working Environments), IIEP (Interchange, Information, Studies and Research), ABC Chemists' Union, Osasco Metalworkers' Union, Social Observatory Institute, ENSP/FIOCRUZ (National School of Public Health/Oswaldo Cruz Foundation) and CEREST/SP (São Paulo State Workers' Health Reference Centre). The title of the magazine analysed is Nanotechnology: Wonders and Uncertainties in the Universe of Chemistry (figure 05).

Figure 05: Nanotechnology Comic Strip Magazine
Source: Jensen, 2010

The context of the comic is the dialogues between the employees of the company Transportadora Novo Universo, a leader in the delivery of dangerous, toxic, contaminating and explosive materials. The dialogues reveal the key concepts of the nano world, such as the nanometric scale; the new properties that arise from materials when worked on the nanoscale (figure 06); the main production processes for nanocomposites in the nanotechnology industries, such as the manipulation of matter from "Bottom to Top", "Top to Bottom" and the process of Self-Assembly.

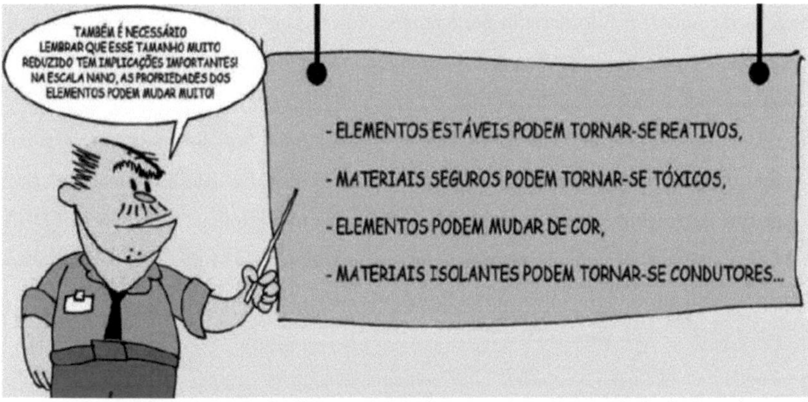

Figure 06: **Excerpt from the magazine "Nanotechnologies: marvels and uncertainties in the universe of chemistry" on the** conceptual principle of the nanometre scale
Source: Jensen, 2010

When it comes to understanding scales, the teacher always uses something comparable to establish a size relationship - this is the case, for example, with distances and heights. However, when it comes to the nanometre scale, this relationship is not always easy and "seems to be a major aggravating factor in understanding this new technology and, consequently, the phenomena involved in it. However, learning and understanding the subject of nanotechnology, as well as its applications, is fundamental in the school environment" (SANNOMIYA, HONÓRIO e PEREIRA, p. 74, 2010). And one of the most didactic ways to overcome this obstacle is to compare the nanometre with some material that is also small but visible, such as the diameter of a hair (figure 07).

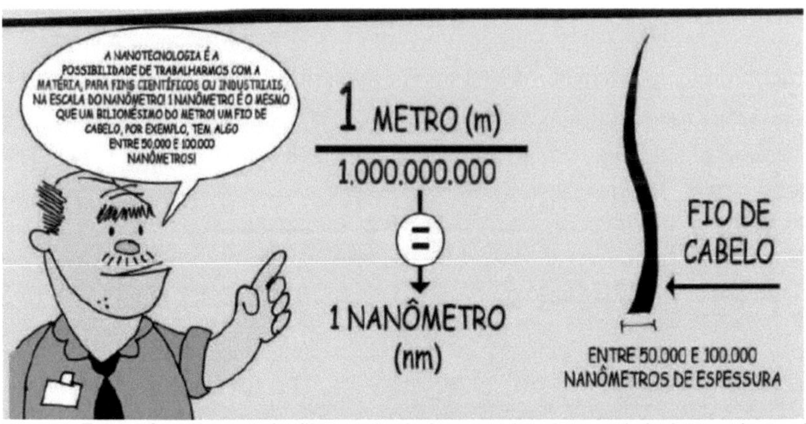

Figure 07: **Excerpt from the magazine "Nanotechnologies: marvels and uncertainties in the universe of chemistry" about the** main characteristics and properties of nanomaterials
Source: Jensen, 2010

Perhaps one of the most impressive characteristics of nanomaterials is that they change their physical and chemical properties, and this can also be misunderstood by students. If a material is divided up countless times, theoretically the atoms that make it up are the same, but their properties should change as a result of the decrease in these atoms. However, this thought is not entirely true and "even worse is being able to show that if the 'little piece' is too small, new properties can appear instead of disappearing" (SCHULZ, p. 05, 2007).

According to Schulz (2009) the "tool-concepts" related to the "top-down" and "*bottom-up*" processes in the production of nanomaterials (figure 08) are intrinsically interdisciplinary because their study requires microscopy, which originated in physics, the transcription mechanisms of DNA molecules in the self-arrangement mechanisms explored in biology and the synthesis of nanoparticles and functionalisation of molecules, which comes from chemistry.

Figure 08: Excerpt from the magazine "Nanotechnologies: marvels and uncertainties in the world of chemistry" about the top-down and bottom-up production process

Source: Jensen, 2010

Understanding a chemical property known as chemical affinity is very useful for the process of producing nanomaterials by self-assembly (figure 09). Molecular associations in this process are an efficient way of keeping nanosystems agglomerated and organised with the ability to perform some characteristic function. Without this chemical linkage it would be impossible to assemble molecular devices one by one on a commercial scale, even with the machines imagined by Eric Drexler in the 1980s (TOMA, 2005).

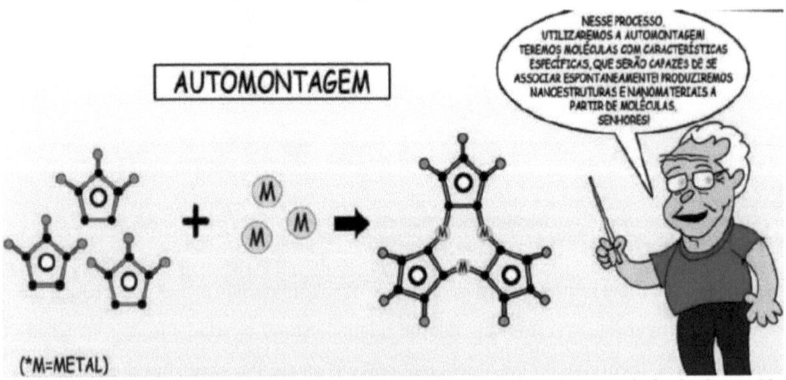

Figure 09: Excerpt from the magazine "Nanotechnologies: marvels and uncertainties in the world of chemistry"

on the production process by self-assembly
Source: Jensen, 2010

The analysis of **the** comic book **"Nanotechnology: Wonders and Uncertainties in the Universe of Chemistry"** shows that there is a wealth of scientific information on nanotechnology contextualised to the petrochemical industry for the production of plastics and derived materials, highlighting its importance for the industrial sector. The contributions of this comic book to chemistry teaching can be seen when it addresses concepts related to atoms and molecules, the transformations that matter undergoes in industry, the reduction of matter that increases the surface area and consequently chemical reactions are altered, among others.

We can also see that the authors were concerned to explain to their target audience the possible risks of environmental impacts associated with the by-products generated by nanotechnology industries, when Olga says: "Actually, I think it's worrying to produce nanocomposites on an industrial scale. Studies on the impact of nanoparticles on the health of workers and consumers - and on the environment, you see, Mr Antônio - are still very limited..." (JENSEN, 2010, p.17).

CHAPTER 3

METHODOLOGY

This chapter is aimed at understanding the methodological steps used in the research in order to systematically clarify each stage of the investigation process. To this end, it is subdivided into the following parts: methodological approach, highlighting the qualitative nature of the case study research; description of the research subjects and the research context; and finally, data collection and analysis instruments, focusing on the observations and on-site records that guided the research.

3.1 METHODOLOGICAL APPROACH

This research is essentially qualitative in nature. Qualitative research has gained a lot of prominence in educational research because it has descriptive aspects and similarities to social research (GIL, 2009).

According to Ludke and André (1986), qualitative research is an activity in which the researcher compares data, evidence, collected information and accumulated theoretical knowledge on a given subject. "It is thus a privileged opportunity, bringing together the thought and action of a person, or a group, in an endeavour to develop knowledge of aspects of reality that should serve to compose proposed solutions to their problems" (LUDKE and ANDRE, 1989, p.02).

In this type of research, the knowledge of education and research methods is shared with the scientific community, however, there is a certain subjectivity in the results because the research subject is endowed with a singularity, so no matter how much you want to eliminate subjectivity completely (STAKE, 1983).

In order to highlight the fundamentals of qualitative research, Bogdan and Biklen (1994) sought out the basic characteristics they pointed out about this type of investigation, namely: the data is collected in the natural environment, with the researcher being the main research instrument; the investigation is predominantly descriptive; the process by which the research takes place is more important than simply the results; the "meaning" is predominantly highlighted for the researcher's analysis; the data is analysed using the inductive process.

Despite characterising qualitative research in five basic points, Bogdan and Biklen (1994) point out that in naturalistic research it is not always possible to follow these points simultaneously, and at one time or another there is greater evidence of certain characteristics.

It is from this perspective that educational research fits in, being a continuous process of transformation of the subjects and of the results, most of the time over the long term. Among the characteristics highlighted by Bogdan and Biklen (1994), it is worth emphasising that regardless of the mechanisms used by the researchers to conduct the research, qualitative investigations focus on the various aspects of educational life.

In education research, modern qualitative approaches have gained prominence among researchers, such as participant research, action research, ethnographic research and case studies, which distance themselves from more traditional approaches without losing scientific rigour. Thus:

> We feel that at the root of current trends in educational research is a legitimate and ultimately dominant concern with the problems of teaching. This is where the roots of the problems lie, which certainly have repercussions on all other aspects of education in our country. This is where research must strike most directly, seeking to make the contribution it has always made to education (LUDKE and ANDRÉ, 1989, p. 08).

Based on these characteristics, this research is part of a qualitative approach in which we sought to understand all the aspects pertinent to the object of study, highlighting the characteristics of the research subjects and placing greater emphasis on the meanings of the data collected. In order to better situate the methodological choice, it is worth emphasising that we opted for a qualitative case study approach.

This type of research originated in the biomedical and social sciences with a descriptive, in-depth and detailed analysis of an individual case to explain the dynamics and pathology of certain rare syndromes. It is a variant of the Problem-Based Learning method, which is based on putting the researcher in contact with real problems in order to stimulate the development of critical thinking, problem-solving skills and the assimilation of concepts (GIL, 2010; SÁ and QUEIROZ, 2010).

From this perspective, qualitative research can be consistently related to the case study, given that:

> A case study is a study of a case, whether it is simple and specific, such as a competent teacher in a public school, or complex and abstract, such as literacy classes or evening classes. The case is always well delimited and its contours must be clearly defined during the course of the study. The case may be similar to others, but it is at the same time distinct because it has its own unique interest. [...]. The interest, therefore, centres on what is unique, what is particular, even if certain similarities with other cases or situations become evident later on. When we want to study something unique, something that has value in itself, we should choose a case study. [...]. The qualitative study [...] takes place in a natural situation, is rich in descriptive data, has an open and flexible plan and focuses on reality in a complex and contextualised way (LUDKE and ANDRÉ, 1989, p. 17-18).

According to Gil (2009, p. 57), case study research "is characterised by the in-depth study of one or a few objects, so as to allow a broad and detailed knowledge of them, a task that is practically impossible with other types of design".

In the field of education, the case study gains new scope and depth due to the complexity of the relationships between the subjects/investigators/process. In this way, Ludke and André (1989) establish the fundamental characteristics of the case study, namely: they aim for discovery; they emphasise interpretation in context; they seek to portray reality in completeness and depth; they use diverse sources of information; they reveal vicarious experience and allow naturalistic generalisations; they represent different and sometimes conflicting points of view; the results use a more accessible language and form compared to other types of research.

For Stake (1983), the contributions of case studies to research into the educational system show that this approach is promising for investigating problems in science teaching, to the extent that researchers explore its positive and negative aspects. The author also reveals that the results of his investigations into the US education system in 1978 would not have had such repercussions if he had used the research and programme evaluation methods employed in the 60s.

According to Ludke (1983, p.15), Stake's investigations using case studies "have arisen at an opportune moment, seeking to respond to the need for better and more effective research contributions to the reality of educational practice, [...] without relying too much on the analytical artifices used by other more traditional methods of scientific investigation". In this way, we can see the importance of this type of naturalistic investigation for understanding the problems faced by the current education system in Brazil, whether in basic education or in teacher training courses for working in basic education.

3.2 RESEARCH SUBJECTS AND CONTEXT

The research subjects comprise a 6th period class of the Full Degree in Chemistry course at the Federal Institute of Rio Grande do Norte (IFRN) Pau dos Ferros campus. There were 19 chemistry undergraduates with very heterogeneous characteristics who participated in the activities developed. The research took place during the Supervised Internship Seminar, one of the aims of which is to consolidate and articulate the knowledge developed during the course through theoretical and/or practical training activities.

Another characteristic of the class, and of the IFRN Pau dos Ferros campus in general, is the difference between the cities of the graduates. According to a previous characterisation of the class, they are on average 22 years old and come from the following towns: São Miguel - RN, Encanto - RN, Itaú - RN, Pilões - RN, Portalegre - RN, Pau dos Ferros - RN, Marcelino Vieira - RN, Pereiro - CE and Uiraúna - PB. This may explain why some of the undergraduates sometimes arrived late because they depended on school transport from their municipalities, which are kilometres away, or forced the teachers to "wait a little longer" to start the lesson.

This characteristic of concentrating students from several different cities and states is due to the IFRN Pau dos Ferros campus being located in a strategic location close to the neighbouring states of Ceará and Paraíba. In addition, Pau dos Ferros is home to the largest number of higher education institutions in the Alto-Oeste Potiguar region, making it a university hub for many students from surrounding towns.

The Federal Institute of Rio Grande do Norte was established in accordance with the provisions of Law 11.892 of 29 December 2008, in accordance with article 5 and item XXVIII, which deals with the creation of Federal Institutes. One of its objectives is to provide higher education with "degree courses, as well as special pedagogical training programmes, with a view to training teachers for basic education, especially in the areas of science and mathematics, and for professional education" (Art. 7°, item VI and (b)).

Of the 29 undergraduate courses offered by the IFRN, 10 are bachelor's degrees, with emphasis on teacher training courses in natural and exact sciences such as Biology, Physics, Mathematics, Nature Sciences and Mathematics and Chemistry. In addition to these, the IFRN offers degrees in Spanish in the classroom and distance learning modality, Geography, Computer Science and Pedagogical Teacher Training for Professional Education. The degree in chemistry is offered by the following authorised campuses: Apodi, Currais Novos, Ipanguaçu and Pau dos Ferros, lasting four years and with a workload of 3,404 hours spread over 8 semesters, including the hours allocated to the internship in basic education.

From this perspective, this research is based on investigating the initial training of chemistry teachers and aims to answer the following research question: what are the theoretical and didactic parameters that teachers in initial training plan and implement in chemistry classes, addressing the content of nanotechnology in a multi- and interdisciplinary way? It analyses the inclusion of current and relevant themes for society in basic education, considering that students need to have a critical and solid education in the development of science and technology.

According to Afonso (2011, p. 27) there are no records of systematic approaches to the teaching of nanotechnology in primary and secondary schools, revealing that "education in Brazil, despite all the progress made in recent decades, faces difficulties in basic aspects such as infrastructure, teacher qualifications, students' living conditions and their motivation to learn". For Afonso (2011), being able to systematise teaching in these conditions seems utopian, but in the context of a country that is growing in scientific and technological development like Brazil, it becomes a cultural and economic necessity.

In the context of the initial training of science teachers, McDermott (1990) apud Carvalho and Gil-Pérez, 2011, p. 69, establishes some characteristics that constitute serious impediments in teacher training, such as:

> I - The lecture format encourages passive learning; future teachers become more

> accustomed to receiving knowledge than helping to generate it. II - The "standard problems" posed lead to somewhat rhythmic, repetitive questions, without contributing to the development of the forms of reasoning needed to deal with new situations, such as unanticipated questions that students might ask. III - Laboratory practicals use sophisticated equipment that is not available in secondary schools and, above all, they are limited to a recipe-style process of verification, which does not contribute at all to an understanding of scientific activity. IV - The breadth of the curriculum covered and the short amount of time devoted to the different subjects prevent an in-depth understanding of the concepts involved, and even less so - we would add - the treatment of aspects such as the interactions between Science/Technology/Society, etc., which are essential for giving a correct image of Science (McDermott, 1990 apud CARVALHO and GIL-PÉREZ, 2011, p. 69).

It is therefore necessary to rethink initial teacher training programmes from a more comprehensive but less generalist perspective, as we seek to intensify the part of the curriculum that encourages investigative practice to the detriment of the passive expository format. For Libâneo (2011, p. 83), teaching as a critical-reflective activity is an indispensable prerequisite for solid initial teacher training. Thus, "it is certain that quality general training for students depends on quality training for teachers".

Schulz (2009) argues, for example, that the teaching of nanotechnology should follow a model that presents scientific research in three ways: driven by context (social, economic, environmental...); focussed on a generally complex problem; and interdisciplinary. These ideas can thus be associated with McDermott's (1990) conceptions and Carvalho and Gil-Pérez's (2011) investigative CTS approach as strong allies for critical-reflective initial teacher training.

3.3 DATA COLLECTION AND ANALYSIS TOOLS: DESIGN OF THE CHOSEN CASE STUDY

The basic structure of well-conducted case studies includes three types of data collection instruments: observation, document analysis and interviews (GIL, 2010; LÚDKE E ANDRÉ, 1986; YIN, 2001). The researcher takes into account naturalistic research factors such as the context, the subjects being researched, the implications of their research for the community and the time available to determine which of these instruments to use during their investigation. It is therefore necessary to briefly discuss the main characteristics of these instruments used in case study research in order to substantiate the choices made in this research.

According to Gil (2010), observation can take three forms: spontaneous, systematic and participant. In the first modality, the researcher remains uninvolved with the group or community they intend to research and stands out by observing the facts that occur in the environment. Spontaneous observation is suitable for exploratory studies in which the closeness between the researcher and the observed phenomenon is favoured. In the second modality, systematic observation, the researcher is no longer oblivious to the group they intend to study; on the contrary, they know which aspects of the group are significant for achieving the intended objectives. This method is suitable for descriptive studies and the researcher is able to draw up an observation plan to guide data collection, analysis and interpretation. Finally, in participant observation, the researcher actually takes part in the group, to some extent being an integral part of the group.

According to Liidke and André (1986, p. 26) "observation enables the researcher to have personal and close contact with the phenomenon being researched, which has a number of advantages". This is why observation has a prominent place in educational research approaches, with introspection and personal reflection being emphasised in this type of qualitative research.

For Yin (2001), direct observation takes the form of visits to the location chosen for the case study. Direct contact with the subjects allows the researcher to assume that the phenomena of interest are not purely historical in nature, but rather to observe behaviours or environmental conditions that are relevant to the research. In other words, the importance of direct observation is that it is a privileged moment that allows the observer to set parameters that point to the most coherent paths in the elaboration and discussion of the data. This approach therefore points to the quality and importance of the researcher's work for the scientific and academic community.

Analysing documents is essential for case study research, since all sectors and bodies have access to and make available collections of documents about a particular group, company, school, university, among others (GIL, 2010). In documentary analysis, Ludke and André (1986, p.38) reveal that these documents "range from laws and regulations, norms, opinions, letters, memos, personal diaries, autobiographies, newspapers, magazines, speeches, radio and television programme scripts to books, statistics and school archives".

According to Yin (2001, p. 109), when using documentary analysis, the researcher "does not rely on the necessary accuracy or the perceived absence of biased interpretations". Its contribution to case studies lies in the fact that it corroborates and valorises evidence from other sources. Thus, during this phase of the research, the documents serve to guide the direction of the research and, when necessary, the comparison of information is indispensable.

Finally, there is the interview, which is based on collecting information pertinent to the case study in question. Gil (2010) defines some types of interviews, such as: open (ample freedom to answer, but following a sequence guided by the interviewer), guided (formulated and sequenced during the interview), by guidelines (the researcher lists the points of interest) or informal (similar to a conversation). For Ludke and André (1986), the great advantage of the interview lies in the fact that the desired information can be obtained immediately and on a regular basis, with practically any type of informant and on a variety of subjects. "In general, interviews constitute an essential source of evidence for case studies, since most of them deal with human issues" (YIN, 2001, p.l 14).

From this perspective, it should be emphasised that the data presented and discussed in this dissertation was collected through systematic observation, as this is fundamentally descriptive qualitative research. The observations took place during the classes of the Teaching Internship Orientation Seminar I, two 50-minute classes per week. In total, there were fourteen meetings that enabled a satisfactory interaction that forms the basis of this case study.

In addition to the observations, it was necessary to consult normative documents, such as the pedagogical project of the chemistry degree course, approved by Resolution No. 08/2012-CONSUP, and others, whose aim was to establish the importance of initial teacher training courses in the area of science teaching.

3.3.1 Initial contact with the class

In anticipation of the first contact with the class chosen for this case study and considering the impartiality required for the information collected, the undergraduates were asked about their prior knowledge of nanotechnology and its relationship with chemistry teaching. It should be emphasised that in this first contact there was no discussion to influence the students' answers, so the questioning only revealed their prior knowledge in a reliable way, since the subject of nanotechnology is widely covered on TV news programmes, in newspapers and magazines with national circulation and on the internet.

Then there was a collective discussion to listen to the undergraduates' dialogue on this topic. Some developed coherent reasoning, while others, more timid and fearful, decided not to speak. In view of the discussions, it was appropriate to present a documentary on nanotechnology to support the discussion that had begun at the first moment. It was the cover story series produced by TV Cultura.

In the documentary, nanotechnology is seen as an evolution of modern science that enables contemporary man to make advances in medicine, discover new material properties and produce clean energy, among other things. The aim of this activity was to introduce students to nanotechnology, its applications and future prospects in an illustrative and dynamic way that would allow them to interact more during later discussions.

3.3.2 Didactic research in the PCN's and links to nanotechnology

Using the National Curriculum Parameters, the undergraduates analysed the topic of nanotechnology from the perspective of teaching with structuring themes based on the creation of thematic units. The activity began with a reading and discussion of the competences and skills developed in chemistry teaching, as presented in the PCNs. The articulation with nanotechnology took place through discussions around the following questions:

01. After reading the PCN texts, what did you understand by "Structuring Themes"?

02. What relationship did you perceive between the structuring theme and the thematic units?

03. How could the subject of nanotechnology be included in a thematic unit?

The questions mentioned above served as a roadmap for the discussions and subsequent didactic activities developed in the subject. The following didactic investigations were carried out by analysing and surveying the contents of secondary school textbooks that could be linked to nanotechnology. The undergraduates learnt which textbooks the schools in their cities would be using and were able to obtain copies on request.

According to the PCN's, thematic units are produced around a specific structuring theme based on planning that can vary from weeks to a two-month period, i.e. thematic units cannot be too long to enable the teacher to achieve the proposed objectives.

3.3.3 Investigative activity in the classroom

The proposal to teach science through investigative activities presupposes that the student is at the centre of the teaching activity, an indispensable assumption within the constructivist perspective. In this way, the principle of research as a teaching method is validated through its fundamentals, namely: students must be confronted by a problem; then they must come up with hypotheses for solving the problem identified; and finally, they must test their hypotheses.

With this in mind, the undergraduates worked on two investigative activities: the first through reading a scientific text and the other through experimentation. It should be emphasised that regardless of whether they are working on a text or an experiment, investigative activities provide students with greater interaction with knowledge by putting them in the position of a researcher in order to solve a particular problem. This problem can be neither too difficult that students can't identify it, nor too easy that it doesn't require them to adopt a researcher's stance.

The first investigative activity was an analysis of the text "Truck loaded with tonnes of ammonia overturns in Pindorama" reported by G1 (annex A), in the classroom. The undergraduates did a brief individual reading to familiarise themselves with the context. The students then formed discussion groups to identify which problem they saw in the accident reported in the text. In addition to the vegetation, there was a river very close to the site of the accident and if the spilled material wasn't contained, it could lead to an even greater catastrophe for the river's aquatic life.

The aim of this activity was to show undergraduates how to carry out research in the classroom using media texts. The conclusion is that it is possible to work on research within any subject, as long as there is some connection between the content being worked on in the classroom and the social and environmental context of the students. Nanotechnology can bring this investigative aspect that problematises social life and the environment into many texts, it's just up to the teacher to direct this focus.

In this investigative activity, each stage was followed by socialisation and discussion in the larger group (all the students and groups). The first group presented their ideas and had the opportunity to defend them through discussion with everyone present, including the teacher, who at this point mediated the discussions. This was followed by all the groups socialising their ideas until a consensus was reached on solving the problem identified.

The second activity was an investigative experiment on the depollution of "water contaminated by textile dyes" (Annex B) using nanotechnological materials such as activated carbon and clays, which have a large specific surface area and great adsorption power.

To this end, the undergraduates made use of the institute's laboratory, but it is worth emphasising that the activities developed do not necessarily require an equipped laboratory, and it is possible to carry out the activity in the classroom itself, as is the reality in most public schools.

The research activity began with the contextualisation that led to the identification of the problem by the undergraduates: the pollution of effluents by textile industries. According to this context, the students were asked to work with a group of researchers hired to re-establish the balance of an aquatic environment near an industry that manufactures dyed clothes.

The materials available in the laboratory were: charcoal, activated carbon, bentonite clay and vermiculite clay. To make up the contaminated water, a solution of methylene blue was prepared, an organic dye that can even be used to dye clothes. The researchers' task was to develop a product with a specific usage strategy that would remove the dye from the water. After discussing their hypotheses in groups of four, the undergraduates proceeded to test their hypotheses by putting the experiments into practice.

After the experiments were completed, there was a more in-depth discussion of the main content developed in that lesson, such as the phenomenon of adsorption, the surface area of materials, the production of activated charcoal in the laboratory and the possibilities of this type of investigative experimental activity in secondary education.

As this is an investigation based on the training of chemistry teachers, the main aim of all the activities carried out was to provide the undergraduates with moments to discuss their applications in real classroom situations.

Therefore, we always resorted to expressions like: How would you work on this activity with high school students during your internship? Imagine your students' difficulties and find a way to make it easier for them to learn the concepts you're working on! Sensitivity to their students' difficulties and the desire to contribute to their education motivates teachers to reflect on their own teaching practices.

CHAPTER 4

ANALYSING AND UNDERSTANDING THE RESULTS

This chapter is aimed at analysing and understanding the results obtained in this case study developed by qualitative research. To this end, it is divided into sections that are linked to the theme of nanotechnology and initial teacher training, namely: the undergraduates' prior knowledge of nanotechnology; getting to know nanotechnology and its applications from a documentary; relating the teaching of nanotechnology to the contents of chemistry classes; research as an educational principle: a selection of the investigative activities carried out; the motivation needed to include nanotechnology in chemistry classes.

4.1 UNDERGRADUATES' PRIOR KNOWLEDGE OF NANOTECHNOLOGY

In qualitative research, the perception of the subjects investigated about the research object is essential for understanding the results obtained. By articulating their world or prior knowledge, the undergraduates can present their initial ideas on the subject of nanotechnology. This knowledge can be considered a product of the subject's conceptions of the world, through interactions established between them and the environment in a sensory, affective and cognitive way, or even as a result of their cultural beliefs (TEIXEIRA and SOBRAL, 2010).

From this perspective, we also looked to the Curriculum Guidelines for Secondary Education to understand that knowledge of chemistry should favour cognitive aspects of undergraduates' understanding of the contents of this subject's curriculum. Contextualisation, in turn, is related to the relevance of the content taught by the teacher to the students' lives, such as the observance of themes like nanotechnology.

The contextual aspects of this technology range from the relationship with chemistry content, through the development of scientific knowledge, to the applications and implications for society. Thus:

> Whatever the chemical knowledge and how it is included in the teaching-learning process, the common basis of the curriculum and the development of basic training competences must be guaranteed. Contextualisation based on socio-cultural aspects is expected, as well as explaining the interrelationships between chemistry, technology, society and the environment, even within the scope of the diversified part of the curriculum proposal (BRASIL, 2006, p. 122).

According to the Curriculum Guidelines for Secondary Education in the area of Natural Sciences and Maths, nanotechnology is seen as a relevant, current issue with strong social relations that deserves attention. According to the National Curriculum Guidelines for Basic Education (DCNEB), "scientific knowledge and new technologies are increasingly becoming a condition for people to know how to position themselves in the face of processes and innovations that affect them" (DCNEB, 2013, p. 26). Thus, the importance of including proposals on how to address issues related to the development of science and new technologies during their professional training in basic education is confirmed. In addition to being multi- and interdisciplinary, nanotechnology arouses curiosity and motivates students by providing new challenges for modern science. Following this context of investigation, the undergraduates surveyed answered the following question: According to your previous knowledge, describe what you understand about nanotechnology, associating it with the knowledge of chemistry acquired during high school, undergraduate studies or even in the mass media (television, internet, magazines, books). Justify your view on the use of nanotechnology as a tool for a better understanding of scientific and technological development in chemistry teaching. Table 05 shows the description of the chemistry undergraduates' prior knowledge on the subject covered in the above question. By organising the data on their prior knowledge, it was possible to identify two distinct categories that facilitated understanding of the undergraduates' initial conceptions of nanotechnology. In category I, there was a group that associated nanotechnology with the scientific and technological development of modern society; in category II, there was a group of undergraduates who identified nanotechnology as a methodological approach or tool used in the teaching and learning of chemistry.

Table 05- Category I: nanotechnology as a scientific and technological product of modern society

	GI	GII	GIII
Nanotechnology as scientific and technological development	*"Nanotechnology is the study of the manipulation of matter on a very small (molecular) scale. This type of technology is still promising, but it could contribute to the development of certain areas such as medicine, chemistry, physics, etc. When it comes to knowledge of chemistry, nanotechnology can help to understand the behaviour of atoms and molecules in the most diverse situations, helping to create new hypotheses and formulate new theories."*	*"Nanotechnology is technology developed at a nanometre level, around 10m. It can be applied in various areas such as IT, medicine and others."*	*"As I see it, nanotechnology is used to create and exploit materials on the nanometre (nm) scale. And from nanotechnology there is the possibility of obtaining new electronic devices that are increasingly smaller ".*

Source: produced by the author

Analysing the data in table 05, it can be seen that a group of undergraduates, using their previous conceptions and knowledge of the world as chemistry undergraduates, managed to understand the general aspects of nanotechnology, associating them with scientific and technological development. Some of them reported that the very morphology of the word "NANOTECHNOLOGY" refers to something quite small, in fact of the order of nano, a Greek prefix meaning dwarf. The term technology, on the other hand, refers to applications in various branches of science, such as medicine, materials science, physics, experimental and computational chemistry, molecular biology, among others.

However, this group was unable to assimilate the use of the study of nanotechnology as a tool or methodological strategy for teaching chemistry, especially with regard to the evolution of scientific knowledge. In other words, it is understood that by contextualising the production of scientific knowledge, nanotechnology refers to the evolution of scientific knowledge, which is called modern science. In this way, students can realise that science has evolved to the point where man can manipulate matter in astronomically small dimensions, and as a result, we can obtain new ways of manipulating matter.
material properties and/or improve existing ones. This leap in science can be seen in the various applications of nanotechnology.

Medeiros and Mattoso (2006, p. 19) make a comparison between the students' previous conceptions and a concept that is well accepted by researchers in the field, stating that "the prefix nano *d* derives from the Greek word *nános*, which means dwarf. In the modern conception of this word, nano is a technical term used for any unit of measurement, meaning one billionth of that unit." To give a chemically conceivable ideal, a nanometre is equivalent to the distance occupied by around 5 to 10 atoms lined up in a row (LAMPTON, 1994).

The idea that "*nanotechnology can help to understand the behaviour of atoms and molecules in the most diverse situations, favouring the creation of new hypotheses and the formulation of new theories*" falls into a conception with a few pitfalls. The recognition of chemistry as a scientific field probably dates back to the 16th century, while nanotechnology only emerged at the end of the 20th century. Much is known about the contributions of chemistry to the development of this new technology; however, even though we know that theories about the behaviour of atoms and molecules can change drastically in a certain period of time (CHASSOT, 2015), affirming this great achievement of nanotechnology should raise some questions.

These questions should be made explicit in initial teacher training, because the scientific education of their students in basic education depends on them. The historical processes of evolution and scientific revolution can help us understand the contributions of chemistry to the development of new technologies, and are therefore seen as essential in teacher training.

For Bernal (1969), research into the origin and evolution of science is complex because it goes back a long way, and the historian, when looking into it, gets very far into the tangled and confusing conflicts of ideas behind science. However, for the development of man and the creation of society as it is, it is necessary to look at how society and science developed at different times in history.

The evolution of scientific knowledge and technologies has been opening doors and stimulating more and more researchers and scientists, as Sagan (1996, p. 281) says: "Science is an astonishment and a pleasure. Every time a spaceship passes over a new world, I am amazed".

Understanding how science has evolved and its direct implications for society highlights the long journey that man has had in establishing himself as a social being who interacts with nature, who produces new technologies and is always in search of the unattainable. Perhaps it is for this reason, man's unattainable quest to overcome his limits, that new technologies emerge with the aim of improving society's quality of life.

From this perspective, Sagan (1996, p. 283) refers to nanotechnology as the "technology for displacing individual atoms, so that long, complex messages can be written on an ultramicroscopic scale. It is also possible to make machines the size of molecules. [...] Where will this take us in a few more decades?". What Cari Sagan treated as a new technology with enormous potential for applications that have yet to be explored, today, almost two decades later, can be considered the technology of the future in a well-explored universe, but with a lot of work ahead for science in the 21st century.

This historical view of science implies a solid training of teachers to work in basic education with reflective practices on the changing nature of scientific knowledge. In a modern society such as the 21st century, school and academic training presupposes critical and reflective training in an attempt to assimilate new knowledge arising from the development of such a society.

In a different vein of thought, category II presents the previous knowledge of the undergraduates about the object of investigation. In this category, a group of students correlated nanotechnology with chemistry teaching methods or tools/strategies for teaching the disciplinary content of this science. This mistaken view of nanotechnology demonstrates a lack not only of knowledge about it, but also of mastery of tools and methods for teaching and learning science.

In the context of initial teacher training, it is worth emphasising that mastery of content teaching strategies is on the same "level" as mastery of the subject to be taught. In other words, at this point, there is a balance between content-based training and the pedagogical foundations already emphasised in this text, which are necessary for the professional practice of any teacher in the natural sciences.

Table 06 therefore shows category II with the chemistry undergraduates' prior knowledge of nanotechnology, considering it as a method of teaching and learning chemistry.

Chart 06- Category II: nanotechnology as a methodological tool or strategy in chemistry teaching

	GI	GII	GIII

Nanotechnology as a methodological strategy	*"The study of nanotechnology is based on the presentation-representation of microscopic molecules, it can be a powerful tool in teaching since chemistry as a science of transformation seeks to emphasise and explore the constitution of matter, however it can be flawed when it is used as the main method of teaching. teaching".*	*"Nanotechnology is a great tool for improving the understanding of scientific and technological development in chemistry teaching, because technology makes chemistry teaching easy and accessible."*	*"Nanotechnology for me is a relatively new area, I have no knowledge about it. I believe it's the involvement of technology to facilitate learning."*

Source: produced by the author

From the analysis of category II presented in table 06 above, it can be seen that a group of undergraduates assimilated nanotechnology as a methodological approach to teaching chemistry, such as a method that would facilitate *the coordination of scientific-technological development in the teaching of chemistry, because technology makes the teaching of chemistry easy and accessible"*. What's more, "[...] it is *a powerful teaching tool since chemistry as a science of transformation seeks to emphasise and explore the constitution of matter, but it can fail when it is used as the main teaching method*".

It should be emphasised that science teaching, in general, has methods and tools/strategies that are already well-established among educators and scholars of the psychology of learning. The most favoured method is in fact the scientific method through induction and/or deduction. From this perspective, within teaching strategies, investigation as an educational principle is the closest to the scientific method (BELLUCO and CARVALHO, 2014; ZÔMPERO and LABURÚ, 2011).

Experimentation followed this same purpose, being incorporated into the teaching axes of biology, physics and chemistry, and thus took on scientific methodology as a teaching methodology (HODSON, 1985). As for teaching methods, the following also stand out

the empiricist rediscovery of science. According to Licillini and Sicca (1992, p. 38) "this method assumes that knowledge comes from observation, which is objective, and that scientific laws come from the induction of facts and data. In this sense, those who use it encourage students to explore phenomena". In this way, the learning of scientifically accepted concepts is based on observation.

However, in a less conservative and more up-to-date view, there is a defence of the freedom of varied pedagogical concepts and the use of teaching tools/strategies, as long as the teacher places them in a critical and reflective perspective. In addition to research and experimentation, the teaching of nanotechnology in chemistry lessons can take the form of documentaries (audiovisual media), discussions or the creation of comic strips, exhibitions of nano-art on screens, concept maps, among others, which have already been defended in other chapters.

Therefore, in this dissertation, the idea of nanotechnology being seen as a teaching method is not conceivable, because it really isn't. It can be considered as multidisciplinary content and very contextualised. It can be considered as multidisciplinary and highly contextualised content in which the teacher, using methodological tools already known in chemistry teaching, can work on the contents of the curriculum interrelated to nanotechnology and, consequently, develop an interdisciplinary pedagogical practice that is contextualised with scientific and technological development.

Some distortions and/or inconsistencies in the graduate students' answers reveal some weaknesses in their understanding and mastery of science teaching and learning methods and tools, which may be reflected negatively during their internships and work as professionals in basic education. If these undergraduates are unable to differentiate technology that is the result of scientific and technological development from chemistry teaching methodologies, at some point in their careers they will stop at merely transmitting information from textbooks in the most traditional way possible.

Still according to the question about nanotechnology and its relationship with chemistry teaching, it was evident that approximately 47 per cent of the undergraduates investigated were unable to formulate a coherent response on the subject, i.e. almost half the class had difficulty talking about something intrinsic to their training.

Some even reported in the questionnaire that they had never heard of nanotechnology, or that it was a new term in their vocabulary. This shows a concern and instigates research in chemistry teaching to develop training work with these future teachers. Nanotechnology is very topical and could be investigated as a practical teaching activity with your students.

But how can these activities be developed if trainee teachers were not encouraged to learn about nanotechnology during their initial training? Or, what should secondary school students learn about nanotechnology? Just concepts and applications? Or can it be shown to be the fruit of the development of modern science? These are questions that highlight the concern about chemistry education and its relationship with scientific/technological development and the environment, which could lead to research in this direction.

In this context, a concern emerges amid the chaos of a society that is increasingly developed scientifically and technologically, but with a large part of the population lacking access to and knowledge about such developments. According to Scliar-Cabral (2007, p. 198):

> Profound transformations in linguistic communications have been observed over the last 25 years, against the backdrop of accelerated globalisation and non-face-to-face contact between interlocutors via the Internet: the fusion of audio, video and telephony in a single system, the massive use of computers and nanotechnology are examples of the scientific and technological revolution that brings about labour relations determined by who has the most knowledge and is ahead in scientific and

> technological innovation: this imposes new processes in teacher training, implying the reformulation of curricula, subject content and teaching methodology itself (SCLIAR-CABRAL, 2007, p. 198). 198).

We can no longer neglect the scientific education of students in basic education by making it the responsibility of teachers trained without a focus on the influences of science and technology on life in society. From this perspective, Siqueira-Batista et al. (2009) advocate teaching nanotechnology with a Science, Technology and Society (STS) approach by establishing four possible thematic axes: nanomaterials, nanobiotechnology and health, nanotechnology and the environment and nanotechnology, ethics and politics.

Thus, according to the authors, in the thematic axis nanomaterials, the teacher can explore the physical and chemical properties of these nanomaterials, such as conductivity, reactivity, melting temperature, among others. In nanobiotechnology and health, the teacher's focus is on the applications of nanotechnology to curing diseases, such as synthesising vaccines, carrying out nano-surgeries, and the prospects for neuroprotection and neural regeneration in future uses.

The third axis is related to the environment and establishes some concerns about the impacts of nanopollution on the environment and consequently on society and human beings, such as assessing the levels of organic and inorganic pollutants in food, the atmosphere and the aquatic environment. Finally, the nanotechnology, ethics and politics strand enables students and teachers to have a critical and reflective discussion as a synthesis of the first three thematic strands presented.

4.2 NANOTECHNOLOGY DOCUMENTARY: A CLOSER LOOK AT NANOTECHNOLOGY AND ITS APPLICATIONS

After this initial diagnosis of the students' representations of nanotechnology, they were shown a documentary from TV Cultura's cover story series on nanotechnology. In this way, the undergraduates were able to better understand its concept, applications and implications, feeling instigated and motivated by the wide applicability of nanotechnology. In addition to the TV Cultura documentary, other videos on nanotechnology have been cited as teaching activities, such as the documentary produced by the Discovery Channel entitled Fantastic Journey - Through the Human Body in Search of a Cure (BASSOTO, 2011).

In science teaching, the video resource has been considered a motivator for learning potentially significant content for students in basic education (SILVA et al., 2012). In addition to the dynamics proposed by the media used by the teacher, this resource helps to contextualise certain topics that are relevant to that moment in the classroom. In addition, students are immersed in a current culture in which society is increasingly connected to audiovisual media, such as televisions, computers, videos, cinemas, among others (ARROIO and GIORDAN, 2006).

Rescuing or reawakening the educational value of the aforementioned audiovisual media can be an encouraging challenge for the laudable task of educating: seeing knowledge as a bridge that makes social and cultural transformations possible. Rooting in this society the educational culture that learning and knowledge itself are not restricted to the school environment is perhaps the greatest challenge for audiovisual resources.

For Arroio and Giordan (2006, p. 11), the potential of audiovisual resources lies in the fact that they "make it possible to carry out studies of intergalactic universes and penetrate realities of microscopic dimensions. Even the most abstract situations devoid of images can be presented by means of some kind of audiovisual structure". Thus, due to its own language of being able to fantasise and/or make perceptible what seems unreal, a video or documentary makes a subject less abstract and enables students to internalise certain content, as well as stimulating creativity.

Rosa (2000) takes a historical look at the use of such resources in Brazilian education and shows that:

> **In the school environment, the use of image and sound as tools to support teaching dates back to the early days of the development of these media. Whether it's with the first attempts to use radio as a tool for educational and cultural dissemination (just remember the official projects such as the** *Minerva Project,* **for example); or the attempts to introduce these media into schools, either through** *TV Escola,* **the Federal Government's most recent investment in the TV area, or through the use of multimedia tools (through the use of CD-ROMs or Internet access), embedded within the MEC project for the acquisition and implementation of computers in schools. The possibility, previously offered mainly by radio and TV, and now by the Internet, of reaching a large number of pupils, often dispersed over large geographical areas, is a siren song for educational planners (ROSA, 2000, p. 33).**

With this in mind, the undergraduates surveyed were offered a didactic activity using a documentary on nanotechnology, the aim of which was to provide an opportunity to discuss the subject, especially as this was a survey on initial teacher training, and to look at aspects of the organisation of teaching through audiovisual resources.

"Nanotecnologia" (Nanotechnology) is a video/documentary produced by TV Cultura and is part of the collection of the programme matéria de capa (cover story), an educational programme whose aim is to provide the public with knowledge about the development of science and technology. The audiovisual programme lasts 28 minutes and is basically in documentary format with three main narratives.

One of the narratives focuses conceptually on nanotechnology, highlighting the prefix that precedes the word technology (**nano**) as something very small, equivalent in value to 10^{-9}, emphasising the intrinsically human interest in measuring things. The other narrative, present in the documentary, advocates the applications of nanotechnology, which although they are barely perceptible, are very present in students' daily lives. A short film produced by students at Cambridge

University is used in the documentary to illustrate this didactically:

A man's beard grows at five nanometres per second. Oh, surprise surprise, a nanometre is very, very small. Despite its size, you can see the effect of things happening on the nanoscale all around you. It gives the sunset its reddish colour, allows birds to sail and helps lizards cling to trees. We humans are strange creatures. We like to measure things. The highest mountain on earth is over eight thousand metres high; a football pitch is about a hundred metres; I'm about two metres tall when I'm standing up and wearing shoes; ants are a few millimetres long; skin cells are about thirteen micrometres. We can't see things that small. A bacterium is about two micrometres, a virus or fungus is even smaller. And so we come to DNA, the basic code of the molecule. The diameter of DNA is two billion times smaller than a metre. Or, more easily, two nanometres, the unit we use to measure the building blocks of everything.

Finally, in the third narrative, there is a dialogue about the implications that a new technology can have on society and the environment, highlighting the ethical aspects of doing science in technological development. These aspects are essential for discussions in the classroom, as they can reflect the antagonistic side of science when it is used for private profit-making interests by the corporations that hold a large part of the scientific capital. The potential risks posed by nanotechnology are still projections and speculation due to the limited research in this area, but it doesn't take many studies to realise that subparticles from processes in the nanomaterials industry are potentiated, causing nanopollution if they are thrown into the atmosphere without adequate prior treatment.

The interactivity of the resources used in this documentary caught the attention of the undergraduates who, although fanciful, characterised the behaviour of matter in the nanometric world where the laws of physics are totally different from those experienced in the micro and macrometric world, such as figure 10 which shows the presenter on the influence of Van der Waals interactions. "Here, where the particles are between 1nm and 100nm, gravity doesn't matter; it's possible to float and even hang from any surface," says the presenter.

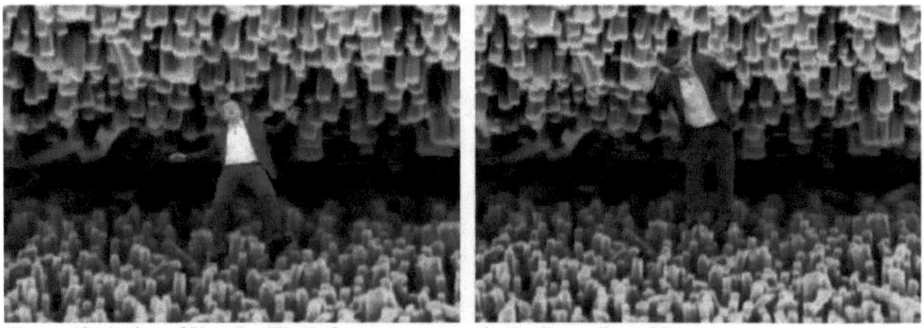

Figure 10: Action of Van der Waals forces on matter in the "nano" world
Source: documentary

Interactivity goes further, as shown in figure 11. In it, the presenter appears sitting on a block of the chemical element called zirconate surrounded by strontium. This is nano-art (coloured screens created using an ultramicroscope, such as a tunnelling microscope) which enchants researchers and observers. Images like this could only come out of man's imagination and become reality after years of research that culminated in electron microscopy techniques.

On a nanometric scale, these methodologies can define and characterise chemical arrangements, sizes and shapes of molecules and particles, as well as study the diversity of structural properties of nanomaterials that are so dependent on atomic size (AZEVEDO, 2006).

Figure 11: Image of strontium zirconate obtained by tunnelling microscopy
Source: documentary

The documentary also shows the multidisciplinary nature of nanotechnology by interviewing researchers from various fields of knowledge, including physics and sociology. When it comes to a new technology, discussions about possible impacts and consequences should be geared towards regulating minimum parameters that establish criteria for research . In other words, strengthening multidisciplinary debates on the subject is a way of liaising between society and researchers, dispelling the often mistaken conception that science is created and discussed only in academic and scientific circles, without the need to listen to society.

As for the applications of the science of the tiny or the invisible, the video/documentary points out that its advances will revolutionise medicine, industry and food production, directly impacting the lives of modern man. Although some applications are still

They are still at the promise stage, but others are already being used and promise to perform small miracles.

In the field of medicine, for example, if something isn't right in the brain, the neurone is replaced with tiny devices capable of performing the same functions as the defective neurone. Nanoparticles could be introduced directly into cancer-causing cells and boost treatments in the battle against cancerous tumours. For Rossi-Bergmann (2008, p.54) the major barrier to overcome when using nanotechnology in health is the rejection of nanomaterials by the immune system, but once this barrier has been overcome, with biocompatible nanomaterials, "nanotechnology is now considered a revolutionary discipline in terms of its enormous potential for solving many health-related problems".

Not only in medicine, but also in cosmetics and dermo-cosmetics, nanotechnology brings fantastic promises, such as the extinction of grey hair, facial creams with nanoparticles that stick better to the skin, shampoos with activated charcoal and zinc nanoparticles, among other promises. According to Gouveia, Garvil and Arantes (2013), the great effectiveness of nanocosmetics lies in their controlled action in the layers of the skin, better penetration of active ingredients into the epidermis and hair, stabilisation of substances and the possibility of controlled release of active ingredients.

In the field of electronic technologies, the video/documentary highlights the development of artificial skin made from silicone and a thin layer of film, over which a solution of carbon nanotubes was applied. Compared to human skin, the artificial skin was resistant, flexible, registered a sensation of touch and did not sag. The function of the carbon nanotubes is that of sensors capable of sensing the pressure exerted on the artificial skin.

According to Teixeira and Salvadori (2007), one of the possible impacts of nanotechnology on human health and the environment is intrinsically linked to the scale of manipulation of materials in nanocomposite production processes. Nanoparticles have sizes equivalent to the pores of cell membranes, which means that they have free access to the immune systems of living beings. If these nanoparticles are toxic, they can have serious implications for society and the environment.

If all this seems like an unattainable promise, you only have to look to the past and realise that man managed to anticipate the scientific and technological revolution that would take place almost half a century later. In this way, the video/documentary takes a historical look at science and technology to show how they have evolved over time, highlighting nanotechnology as the fruit of these developments.

In the 1940s and 1950s, advertisements for various electronic devices such as mobile phones, baby monitors, colour televisions with the option to pause programming, etc. were seen as nothing more than science fiction that stirred up the most sceptical people at the time. Today, thanks to the development of scientific and technological knowledge, it is no longer science fiction, but an outdated reality.

4.3 RELATING THE TEACHING OF NANOTECHNOLOGY TO THE CONTENTS OF CHEMISTRY CLASSES

A lot of research in chemical education has shown the complexity of the teaching and learning process, in an emancipatory conception of the subjects of this process. As a result of this research, researchers in the field of teaching have fully realised that students can no longer be considered passive recipients of the content to be taught by the teacher. Rather, they are subjects with prior knowledge from their culture and social context who need an environment favourable to the culture of scientific knowledge provided by the school (SCHNETZLER E ARAGÃO, 1995).

As a result, research into initial teacher training cannot be carried out without analysing the contributions of research into chemistry teaching, since this is part of a context of reflection and criticism about the professional performance of future teachers. To deny the importance and meaning of teaching research for chemical education is to reproduce a wave of education professionals with conceptions that have lasted for a long time: that in order to teach you only need to know the content, without analysing the pedagogical skills necessary for the teaching and learning process.

According to Zanon et al (2006), this act of reflection on the role of the chemistry teacher in the teaching process starts with an analysis and study of the National Curriculum Parameters (PCNEM) and, consequently, the competences and skills developed or expected in chemistry teaching. In addition, it is necessary to consider that the process of analysing these competences "also presupposes the diversity of pedagogical conceptions, given that these are various theoretical-methodological visions, even if they are interconnected by the opposition to the so-called content of the transmission-reception model" (ZANON et al., 2006, p. 208).

Schnetzker and Aragão (1995) attribute a simplistic view of teaching to the training offered by degree courses, which is often inefficient for the reflective training of teachers. Therefore, an indispensable didactic activity for an aspiring teacher is undoubtedly to look at pedagogical concepts, understand the theoretical-methodological visions and developed competences presented in the National Curriculum Parameters, in the case of a chemistry teacher training course, the chemistry-specific PCNs.

Zanon et al (2006) advocate a balance between teaching and learning models, stressing that one cannot conceive of a teaching model that favours process over content, and recognise the centrality of chemistry content in defining the competences needed to exercise citizenship. There are also groups that propose interdisciplinary teaching, but it is a deductive mistake to presume the end of disciplines in this model.

Generally speaking, the teaching and learning process must be strengthened by balancing these teaching models, and the diversity of pedagogical conceptions is essential for this, since these are different theoretical-methodological visions, although they are interconnected by the opposition of the transmission-reception model of content.

In the context of the research into the initial training of chemistry teachers, the undergraduates surveyed were motivated to investigate the subject of nanotechnology in the light of the competences and skills to be developed in chemistry teaching (Appendix B) with the aim of realising where the teaching of nanotechnology could fit into chemistry lessons. It was therefore necessary to consult the National Curriculum Parameters for Secondary Education (PCNEM) and note that:

> The general competences to be developed in the area of Natural Sciences, Mathematics and their Technologies relate to the domains of representation and communication, involving the reading and interpretation of codes, nomenclature and texts specific to chemistry and science, the transposition between different forms of representation, the search for information, the production and critical analysis of different types of texts; investigation and comprehension, i.e. the use of ideas, concepts, laws, models and scientific procedures associated with this discipline; and socio-cultural contextualisation, i.e. the insertion of disciplinary knowledge in the different sectors of society, its relations with political, economic and social aspects of each era and with contemporary technology and culture (PCNEM, 2000, p. 88). 88).

In the PCNEM guidelines for chemistry, it can be seen that the specific competences for each domain relate to the contextualisation of this science as an indispensable discipline for understanding the physical and natural world, linking it to students' daily lives in the most varied aspects of their lives. It is recurrent to observe what these competences are, in order to awaken the ability to relate them during the teaching of nanotechnology as an essential character to guarantee the effectiveness of learning and meaning in chemical education. These competences are presented in Table 07 (BRASIL, 1998):

Chart 07: Competences and Skills in Chemistry according to the PCNEM

Representation c Communication	
Describe chemical transformations in discursive languages. - Understand the codes and symbols of current chemistry. - Translate discursive language into the symbolic language of chemistry and vice versa. Use the symbolic representation of chemical transformations and recognise their changes over time.	- Translate discursive language into other languages used in chemistry: graphs, tables and mathematical relationships. - Identify sources of information and ways of obtaining information relevant to knowledge of chemistry (books, computers, newspapers, manuals, etc.).
Research and Understanding	
Understanding and using chemical concepts from a macroscopic (logical-empirical) viewpoint. - Understand chemical facts from a macroscopic viewpoint (formal-logical). - Understand quantitative data, estimation and measurements, understand proportional relationships present in chemistry (proportional reasoning). - Recognising trends and relationships from experimental or other data (classification, sorting and correspondence in chemistry).	- Select and use scientific ideas and procedures (laws, theories, models) to solve qualitative and quantitative problems in chemistry, identifying and monitoring the relevant variables. - Recognise or propose the investigation of a problem related to chemistry, selecting relevant experimental procedures. - Develop hypothetical-logical connections that enable predictions to be made about chemical transformations.
Socio-Cultural Contextualisation	

Recognise relevant chemical aspects in the individual and collective interaction of human beings with the environment. - Recognise the role of chemistry in the productive, industrial and rural system.	• Recognise the ethical and moral limits that may be involved in the development of chemistry and technology. • Recognise the relationship between the scientific and technological development of chemistry and socio-political and cultural aspects

Source: Brazil, 1998

One of the teacher's most important tasks is to select and organise content in a systematic and harmonious way in their planning. The PCNEM therefore proposes selecting and organising the content to be taught through structuring themes, "which allow a body of knowledge to be developed in an articulated way, around a central axis with its own objects of study, concepts, languages, skills and procedures" (PCNEM, 2000, p. 93). The following structuring themes are suggested:

I- Recognising and characterising chemical transformations;
II- First models of the constitution of matter;
III- Energy and chemical transformation;
IV- Dynamic aspects of chemical transformations;
V- Chemistry and the atmosphere;
VI- Chemistry and the lithosphere;
VII- Chemistry and the biosphere and
VIII- Quantum models and chemical properties.

Within each structuring theme are thematic units with the contents that make up the subject's curriculum. After the undergraduates analysed the topic of nanotechnology, it became clear that it could not be considered a structuring theme, although structuring themes are not restricted to the topics suggested by the PCNEM, but rather as a multidisciplinary and contextualised theme that cuts across various contents or thematic units in chemistry.

It was therefore perceived that the theme analysed could be included throughout the development of the activities, not in a one-off way, but gradually and integrated into the content through methodological strategies suggested in the literature. For example, it would be possible to work on a video/documentary about nanotechnology and during the teaching of the contents of the following lessons, in the feedbacks, recall and contextualise nanotechnology with scientific and technological development.

In this context, the undergraduates researched the main chemistry contents of the curriculum in the textbooks most commonly used in schools in the region that made direct or indirect links with nanotechnology. After selecting them, they gave their reasons for choosing them. The main chemistry contents identified by the undergraduates are shown in the graph in figure 12 below:

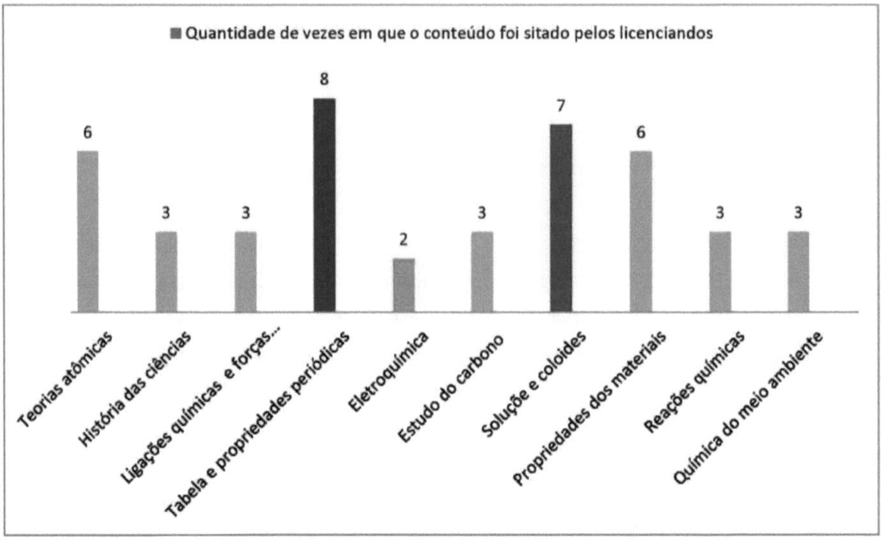

Figure 12: Main contents related to nanotechnology, according to undergraduates
Source: produced by the author

During the discussion and socialisation of the content, there was a time to justify the choice of a particular piece of content, making the connection with nanotechnology. Among the main justifications given by the undergraduates were:

- Historical assumptions of chemistry that demonstrate the history of scientific and technological development in search of human quality of life;
- The periodic table and the periodic properties of the elements, where the most cited was the size of the atoms and ions, which are in the order of nanometres;
- Chemical bonds and intermolecular forces correlating the types of forces and interactions in matter when worked with at nanometre levels;
- The chemistry of carbon, its allotropic forms and the most recent discovery thanks to advances in nanotechnology: carbon nanotubes with semiconductor properties and great storage capacity, even surpassing the silicon used in chips;
- Colloid chemistry emphasises the size of the dispersed particles, which can vary between 1nm-1000nm and consequently determine the type of solution. Nanotechnology works by manipulating atoms and molecules between 1nm-100nm.

In this way, understanding the properties of colloidal systems has made a decisive contribution to the foundations of nanotechnology.

- In electrochemistry, an experimental activity mentioned by some of the undergraduates that immediately made a connection with nanotechnology is electroflocculation. By passing an electric current through an electrolytic solution, colloidal particles are formed and students can see the formation and growth of these particles. Discussions can revolve around the scale of the various particle sizes, from nano to micro and macro.
- In terms of material properties, a comparative study can be made between the main properties of materials on a macroscopic scale and those on a nanoscale; how nanomaterials are produced in the chemical industry; and even examples that are easy to understand, such as the fact that in bar form it doesn't combust, but when finely divided it is used in rocket propulsion because it is very reactive.
- The scope of application of nanotechnology can be related to the high reactivity of compounds on nanostructured surfaces. Thus, chemical reactions that occur at macromolecular levels can be correlated with reactions at nanostructural levels. One example is the dependence of the speed of reactions on the contact surface. Some materials don't react at all or take an enormous amount of time to react because they have a small contact surface.
- With regard to environmental chemistry, discussions can be based on the impacts that the nanomaterial production and use industry can have on the environment, such as nanopollution and intoxication, emphasising the importance of preserving the atmosphere and aquatic life based on scientific research ethics.

Nanotechnology can be taught in chemistry lessons when the teacher can see at least one of the many links between chemical knowledge and this new technology. Therefore, when planning their lessons, teachers should encourage students to investigate and research the subject from a multi- and interdisciplinary perspective so that they can see the links between chemistry content and nanotechnology, as well as the complementary relationship between the disciplines that make up scientific knowledge, given that:

> Dialogue between disciplines is favoured when teachers of the different curricular components focus on the real context as the object of study - the situations in which students live, natural and artificial phenomena, and technological applications. The complexity of these objects requires multidimensional analyses, with the significance of concepts from different conceptual systems, translated into school subjects (BRASIL, p. 102, 2006).

Thus, according to the undergraduates, the teaching of nanotechnology can be related to or included in a thematic unit within the structuring themes, as shown in the following statements:

"In the *2nd year of secondary school we can include the subject of nanotechnology when studying electrochemistry in the thematic unit Aqueous electrolysis, we can work with electroflocculation in order to get an idea of the size of the nanoparticles.*

(Srt)s particles ". (Sic)

"I think one possibility would be to work on the theories and applications that are already suggested in the book, showing what nanotechnology can add to our lives according to the topics that are covered. Example: Electricity: When explaining this subject, you could add that nanotechnology can be used to produce more in a sustainable way. Look: a car is built, but it doesn't have an engine. The energy to move the car comes from painting it, in which a liquid electrolyte has been mixed with the paint, capable of converting light into clean, cheap and completely renewable energy. In this way

(and) try to familiarise ourselves with nanotechnology". (Sic)

"Nanotechnology can be used in a series of moments within the subject of Chemistry, it is important to note that in the face of all the content of chemistry we will deal with microscopic particles, such as the atom, it is in this type of approach that we can use nanotechnology for a more in-depth explanation. We took the historical context of chemistry as our thematic unit, and nanotechnology could be used to explain the history of chemistry.

formation of matter or in Dalton's theory and so on (Sic)

"There are many ways of working with the subject of nanotechnology in a didactic unit. Among them is the experimental activity, which, taking the theme of nanotechnology in water treatment as an example, is a way of working with experimental procedures (Sic).

According to the first statement, nanotechnology would be contextualised with the formation of colloidal particles through electroflocculation by means of an electric discharge. The formation of these particles can be clearly seen without the aid of microscopes and their sizes can be compared and analysed according to their properties, which enables high school students to understand surface phenomena such as adsorption. In other words, as the size of matter decreases, new forces begin to interact more strongly with the environment and nanotechnology is able to take advantage of this very well.

In the second statement above, the graduate showed the relationship between the teaching of nanotechnology and chemistry through new forms of energy production, citing as an example the possibility of a vehicle moving by converting light energy into electrical energy capable of moving the vehicle. This is undoubtedly one of the most promising areas of this new technology: producing renewable, clean energy that is not only efficient but can also help to reduce environmental impacts.

In the third, there is a relationship between the thematic units, but the vision of an epistemological obstacle can hinder the teaching of nanotechnology, since secondary school students need to understand the constitution of matter in terms of energy models. In other words, atoms and small molecules are energy and you can't see them by looking through the lens of a microscope. When students think of the atomic model as being microscopic, they insist on the idea that atoms are visible in laboratories by simply bringing an ordinary microscope closer to a material to be analysed. However, the advances in nanotechnology are also welcome in the manufacture of super-powered microscopes such as tunnelling and scanning microscopes, which allow for the approximate imaging of a few atoms or clusters of atoms and molecules.

In the following statement, we see the relationship between the development of nanotechnology products for water treatment, whether from contaminated rivers and springs or for reuse. Seeing progress in this area is more than just didactic activities in the classroom, it involves raising environmental awareness about the need to look after this indispensable commodity for life, water. One of the main uses of nanotechnological materials in water treatment is activated carbon, which has a high surface area, excellent adsorption sites and good interaction with various chemical contaminants (XIAOLEI QU, ALVAREZ and QILIN LI, 2013).

Tasca et al (2015) developed a didactic experiment involving the preparation and application of superparamagnetic nanoparticles (NPMag) which is in line with the perspective shown by the graduate in statement five. The aim was to show students a practical application of nanotechnology in the treatment of water contaminated by accidents in oil basins.

In the synthesis of NPMags, it is possible to incorporate an organic layer of oleic acid into the nanoparticles so that they can interact satisfactorily with the spilt oil and other organic phases present. With the aid of a magnet, the oil stain removal process is completed by attracting the supermagnetic nanoparticles impregnated with oil and organic components present in the environment.

Modified magnetic materials are currently in the limelight and have been intensively researched by the scientific community due to their properties and potential applications in various areas such as biology, medicine and the environment. These applications include magnetic resonance imaging (a technique used in cancer diagnosis), cancer therapies (drug vehicles and magnetically controlled hyperthermia) and the removal of toxic materials from industrial waste (YAMAURA et al., 2004).

Also with a view to incorporating nanotechnology into the chemistry curriculum, according to the Curriculum Guidelines for Secondary Education in the area of natural sciences and the National Curriculum Guidelines for Basic Education (DCNEB), technological advances and growing scientific development should be worked on in schools with themes such as nanotechnology through methodologies or strategies with a specific aim to develop skills and concepts in this area of science.

The research carried out into nanotechnology can increase concerns, for example, about the environmental impact of oil spills on oil rigs. As these are fairly recent studies, the risks to the environment are not known for certain. Therefore, in parallel with these studies, research should evaluate and seek to develop mechanisms that minimise the impact on the environment. In this way, discussions are inserted into practices that can lead to Environmental Education within nanotechnology.

Nevertheless, despite the guidelines of the Ministry of Education's normative documents, recent research into scientific dissemination reveals that the teaching of nanotechnology is restricted to activities such as the use of videos or documentaries and debates and reading of scientific dissemination texts integrated into the syllabus of school curricula (PAULINI DE JESUS and HIGA, 2014).

In other words, to date there has been no proposal to link science curricula to nanotechnology and research into science teaching in this area can still be considered incipient. Part of this problem can be found in teacher training courses, where their curricula do not include teaching practices on the applications and implications of new technologies.

Teachers may not yet feel encouraged to break away from the traditional sequence of teaching methods that restricts science content traditionally established over the years. As well as understanding the importance of nanotechnology, these teachers need to be encouraged to work with their students, from an investigative perspective, on the evolution of scientific knowledge and its implications for modern and contemporary society.

The history of science shows how scientific knowledge evolves and materialises in contemporary society, and since nanotechnology can be considered an evolution of science, there is the possibility for teachers to develop dynamic work in chemistry classes on the production of new technologies throughout the history of science. Several scholars point to the need to include the history of science in teacher training curricula and in primary schools, such as Porto (2010):

> Recommendations for the inclusion of the history of science in curricula, and in particular in teacher training curricula, can be observed in various countries - in some cases for decades; in others, the discussion is more recent. It should be remembered that the educational objectives of decades ago are not those of today; and, similarly, conceptions of the history of science itself have changed over time (PORTO, 2010, p.159-160).

From this perspective, it can be seen that the focus given to the history of science in curricula varies greatly according to the educational objectives at different times. In this way, the author describes that there was a time when curricula described the historical production of science as linear, cumulative, made by "geniuses", pointing only to progress, in other words, there was no critical look at how science was produced. This is very different from today's concept, which advocates a critical view, based on science as a human construction, on the historical aspects of its production and inserted in cultural, socio-economic and political contexts (PORTO, 2010).

According to Schulz (2009), science teaching that does not stray from its historicity and does not lose its interdisciplinary focus must be based on the possibility of restructuring teaching and research institutions. To do this, the author uses a theory called "mode-2 of knowledge production" proposed by Michael Gibbons and collaborators (1994), which is established as follows:

> In this mode-2, scientific research would be driven by the context (social, economic, environmental...), focused on the problem (usually complex, such as global warming, for example) and interdisciplinary. It is also common within mode-2 to establish networks of researchers, as is the case with the Genome Project. Mode-1 would be the traditional environment of research universities: based on formal separation into traditional disciplines and often individual initiatives by scientists (SCHULZ, 2009, p. 35-36).

Although there are criticisms of this teaching model proposed in large research centres such as universities, it should be pointed out that its implications for basic education are pertinent because there is still a very strong tendency to copy models that have worked at some point. Until basic education schools find a true profile that coherently consolidates their objectives, they will always need to copy models.

In this way, science teaching cannot neglect the production of new technologies in its curricula, considering their historical aspects and interdisciplinarity, such as nanotechnology. The school thus becomes a place for thinking about the direction of these new technologies that are emerging. According to Toma (2005, p.09) "the issue of nanotechnology cannot fail to be included in the education scenario, since the main players in this new era will be young students, who will need to be prepared to enter this labour market".

4.4 RESEARCH AS AN EDUCATIONAL PRINCIPLE: A SELECTION OF THE INVESTIGATIVE ACTIVITIES CARRIED OUT

The methodology of enquiry teaching tries to narrow the gap between the science taught at school and the science produced by scientists and researchers, showing that the greater this gap, the more difficulty students will have in their scientific education and, consequently, in understanding the world around them. Although it is clear that the level of knowledge required in academic science is high, due to the scientific rigour required in research, students' education cannot be neglected by depriving them of producing their own knowledge through research (ZOMPERO and LABURÚ,

2011; ANDRADE, 2011).

For Munford and Lima (2007), science in these two contexts has different objectives: "the main objective of the school is to promote the learning of centred knowledge that has already been consolidated, while, on the other hand, the main objective of academic science is to produce new scientific knowledge".

It should be emphasised that the availability of material infrastructure and qualified personnel in academic science is much greater than that found in schools, and this creates difficulties in implementing research that performs well. When it comes to human resources, more precisely teacher training, many teachers have not received extensive training to develop this methodology, but research shows that continuing training can make up for this shortfall (MUNFORD and LIMA, 2007).

What's more, this methodological proposal escapes the concept of passivity imposed by the transmission-reception method of traditional teaching, which in no way favours the critical formation of the student or the reflective practice of the teacher. On the other hand, this proposal demonstrates critical and reflective attitudes towards the teacher-student dyad in the teaching and learning process.

In the Brazilian educational context, the discussions surrounding the teaching of science by enquiry should be consistently observed, in order to see to what extent schools and professionals are prepared or are adapting to receive this teaching proposal. However, Munford and Lima (2007) emphasise the following:

> A series of questions can emerge from this kind of panorama. To what extent are the initiatives adopted here organised around a coherent theoretical framework? Have dilemmas and challenges that arise in other contexts been taken into account when drawing up national initiatives in the field of research and teaching? What are the implications for teacher training? These discussions are of great importance in our country, as they explain ways of thinking about enquiry teaching, creating a space for dialogue between researchers and educators involved in science teaching (MUNFORD and LIMA, 2007, p. 03).

Considering this context, the undergraduates worked with a text published on a nationally known news portal. Investigative practice with text in the classroom is based on the simplicity of using resources or media. Although it doesn't have the same dynamics as experimental activities, investigative textual practice provides moments of significant relevance for discussion and argumentation in the construction of hypotheses.

All the undergraduates were able to successfully identify the problem reported in the text (attached): the contamination of the soil and one of the effluents of the São Domingos river by the chemical substance that leaked during an accident, as well as reporting the danger to the people around the place where the accident occurred. The text worked on with the undergraduates was easy to interpret and understand, given that the aim of the activity was to demonstrate the possibilities of transposing it to the primary school classroom, and thus adapt it to the level of secondary school students.

Once the problem had been identified, the arguments were discussed and defended with the whole class. This moment is crucial to the success of the research activity and the teacher's role as mediator of discussions and stimulator for students to argue became clear, demonstrating that there is no room for fear of error or saying something "silly" (CARVALHO, 2013). But rather, in the construction of knowledge, whether at school or in the academic spaces of researchers and scientists, error is essential in order to achieve the most reliable results possible.

According to the text, the chemical substance was the salt "ammonium nitrate", a compound used in the manufacture of explosives, which is why there was concern about isolating the site. It can be seen, then, that some media outlets express themselves in the wrong way when it comes to the nomenclature of certain chemical compounds. On purpose, there was no prior discussion of this "misunderstanding" in the report so that in the process of analysing it, the audience could learn more about it.

the undergraduates could point it out. However, there was no comment on this substance.

In fact, the compound the article is talking about is ammonium nitrate (NH_4NO_3), not ammonia, a salt widely used as a fertiliser because it provides nitrogen to the soil through both the nitrate anion (NO_3^-) and the ammonium cation (NH_4^+). This shows that teachers can explore with their students the falsifiability of data presented on websites, even on nationally recognised channels. Although it is recognised that the nomenclature of chemical compounds is very particular to those in the field, there was no concern to pass on the news correctly, asking someone who understands the subject to proofread and correct errors.

Analysing the hypotheses raised by the undergraduates, it emerges that many reported very superficially on their knowledge of chemistry. The main hypotheses raised were related to isolating the area, removing the spilled material and applying reactive chemical products to absorb the contaminating material.

While it is possible to perceive elements of the content already studied by the undergraduates, for example: as this is an activity for chemistry classes, a very reasonable hypothesis for solving the problem would be to correct the pH of the altered soil. The pH of the soil is normally around 7, which makes the soil close to neutral and favourable for vegetation to grow.

When the ammonium nitrate is spilled, the soil becomes acidic due to the reaction between the ammonium ion and the water present in the soil, requiring the application of a product that reacts to neutralise the acidity of the soil. The following equations describe the equilibrium that makes the soil more acidic, as well as a possible reagent used to neutralise it:

$$NH_4^+ + H_2O \leftrightarrow H_3O^+ + NH_3 \quad (1)$$

$$CaCO_3 + 2\,H_3O^+ \leftrightarrow H_2CO_3 + 2H_2O + Ca^{2+} \quad (2)$$

$$\downarrow$$

$$H_2O + CO_2 \uparrow$$

Nanotechnology was also addressed during an experimental investigative activity conducted by the undergraduates with great enthusiasm. On this occasion, the undergraduates made use of the laboratory to carry out the activity. As the proposal was to integrate a problem that had to be solved by investigation with concepts learnt about nanotechnology in previous discussions, we started with a context that is well known today: the pollution of rivers by the discharge of liquid effluents from textile industries.

After reading, discussing and socialising the problem identified (text attached), the undergraduates tried to come up with hypotheses about how they could work to clean up water that was dirty with textile dyes. In the laboratory, dirty water containing the organic dye methylene blue was simulated and the materials available for research were: charcoal, activated charcoal, bentonite and vermiculite clays, as well as common materials and equipment for stirring and filtration.

The main hypotheses raised include:

"raatrr the polluted water with clay, it absorbs the pigment from the flowing liquid by colloidal reaction; Removal of impurities from the contaminated water by adding activated charcoal; Addition of bentonite clay for adsorption of the dye; Adsorption of the dye through charcoal; Add bentonite clay to the polluted water and then filter with activated charcoal" (Sic).

Both activated carbon and some clays have a large specific surface area and their excellent properties as adsorbents are used to remove organic pollutants and even oil stains when there are leaks on oil platforms. These are two materials that are used in nanotechnology research and are very close to the reality of secondary school students. Coal can be bought commercially or produced in the laboratory under specific conditions (quite expensive), while clays can be found naturally. The adsorption process is the one most often presented in scientific research results on nanoparticles or nanostructured materials, which shows a direct correlation between this phenomenon and nanotechnology (QUINA, 2004).

In this way, the undergraduates were able to see that the contents and main concepts of nanotechnology can be worked on from an investigative perspective through experimentation.

Analysing their hypotheses, they all associated direct contact between the available materials and contaminated water (methylene blue solution). They used some wrong terms, such as absorbs, when in fact it is adsorbs. However, during the discussions it was possible to explore some of the obstacles encountered in order to indicate the most scientifically coherent path. Secondary school pupils have the same obstacles and the teacher must bear in mind that the conceptions that pupils bring to the classroom persist, but that they must be replaced by scientific knowledge in a not very easy process of negotiation and persuasion.

In the next stage (testing hypotheses), the undergraduates used experimentation to bring the materials into contact with contaminated water. At this point they discussed how to proceed in order to obtain more scientifically credible results. For example, they would have to weigh the same amount of material, leave it in contact for the same amount of time, filter it under the same conditions and write down all the observations in a logbook. In this way, they would have an understanding of which factors could or could not interfere with the removal of the dye.

During the practical activity, the undergraduates noticed that some of the materials were unable to remove the dye. At this point, they realised the importance of testing hypotheses during experimentation in order to change their preconceptions. On the other hand, other materials managed to clean the water quite successfully by removing the methylene blue dye. From the results obtained, they were able to conclude that the best results for removing the dye came from activated carbon and bentonite clay, respectively. These results instigated and led to discussions after the practical activity, serving as a basis for other research activities.

The process of producing activated charcoal was highlighted in a very fruitful discussion about the activity carried out, as well as the possibilities for implementing it in primary school classrooms.

Activated carbon can be considered the purest form of carbon. Obtained by physical or chemical activation, its application in the removal of organic dyes in aquatic environments is due to some intrinsic properties, such as: it has large porous structures, relatively high surface area, functional surface groups with great affinity for various adsorbates and the ability to adsorb molecules of varying sizes without modifying their chemical composition (AVELAR et all, 2010).

Due to extensive industrial activity, large volumes of organic dyes used in the dyeing of their products are discharged into bodies of water, causing contamination and consequent extinction of aquatic life, mainly because they inhibit photosynthesis activity by making it impossible for microorganisms to absorb sunlight. Therefore, waste materials are used to obtain activated carbon due to their cost/benefit/efficiency ratio, such as corn cobs and piassava fibres (SALES et all, 2015; AVELAR et all, 2010).

The starting materials for the production of activated carbon are generally rich in fibres and have porous morphological structures. However, when they undergo chemical activation with chemical agents such as zinc chloride ($ZnCl_2$) or phosphoric acid (H_3PO_4), new pores are activated, increasing the surface area and, consequently, adsorption is favoured. These processes involve the manipulation of matter on a nanometre scale, i.e. the pure application of nanotechnology (AVELAR et all, 2010; YAMAMURA, 2009; TOMA, 2005). Figure 13 shows the structures of a porous material in natura and chemically activated with ZnCl2 used to obtain activated carbon:

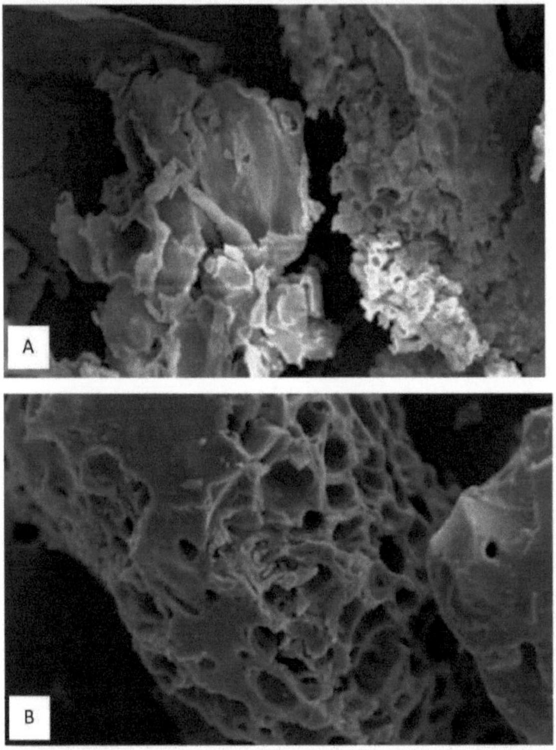

Figure 13: Electron microscopy of raw corncob (A) and corncob activated with ZnCl2 (B)
Source: SALES et all, 2015

By analysing the electron microscopies in Figures 13-a and 13-b, it is possible to clearly see the formation of pores after activation with the chemical agent ZnCl2 and to discuss the increase in contact surface area. It is reasonable to assume that the surface area is proportional to the number of unobstructed pores present in the material analysed and, therefore, chemical activation is one of the most efficient means for this purpose. Discussions can also be based on the use of techniques such as electron and tunnelling microscopy, which took a huge leap forward with the development of nanotechnology at the end of the 20th century. This association between modern techniques for observing matter and nanotechnology can be a very constructive link for students to assimilate fundamental concepts.

Given this contextualisation of nanotechnology and the investigative activities developed, we discussed some ways of using it in practical classroom situations that the undergraduates could address during their internship. The following proposals emerged from the discussions:

"We would *propose a practice of purifying water by adding some materials. Beforehand, we would seek out the students' opinions on the practice, asking which material would be most effective. From there, the practical would begin. At the end of the activity, the students should write down all the data and talk to each other about what really happened, checking if what they had initially thought was correct. Finally, I would ask the students to study a little more about nanotechnology and in the next lesson I would hold another discussion about the experimental activity and conclude by asking* **THE** *students to build a report on the activity." (Sic)*

It can be seen that there was consistency in what the graduate student said, since he proposed an activity based on dialogue between his students and relevant discussions with in-depth research on the subject of nanotechnology. This graduate goes further when he proposes not using a mere script for the experimental practical activity, but rather having the students construct it themselves after the activity. This gives the graduate a high degree of criticism and reflection of teaching practice in terms of the centrality of the action developed by the students in the learning process, which in this case is the construction of their own methodology for solving the problem proposed by the teacher.

"I would use nanotechnology to solve the problem presented. Based on this idea, I would ask how nanotechnology could remove the spilt material in order to *decontaminate the soil" (Sic).*

One of the great advantages of working with the subject of nanotechnology is undoubtedly to demonstrate its applicability in social and scientific life. In this way, the graduate student showed interest in bringing in the experimental investigative activity before starting to delve deeper into the subject under discussion to show some of its applications, and from this, build knowledge and introduce new concepts.

"First of all, I would analyse the content of the Chemistry subject being taught in the class, and using it to create a "bridge" between the subject and nanotechnology. A subject that is taught to students as early as the first year of secondary school and which is part of their daily lives is the content of separating mixtures, and it would be possible to build this bridge with the students by showing that it is possible, for example, to treat polluted water using nanotechnology, since this process uses some methods from the process of separating mixtures" (Sic).

At this point we can see the didactic mastery between content knowledge and pedagogical knowledge in tune with each other, as the graduate realised that one of the teacher's primary tasks is to be guided in the selection of subjects and activities developed in the planning. This shows that not all the content is interrelated with the topic of nanotechnology, but based on this awareness, the graduate promotes a discussion about school planning of the activities developed during lessons so that the moments in the classroom are really productive and encourage students to seek systemic knowledge.

"If the students already had previous knowledge, I would go deeper into that knowledge, talking about the nanometre size, how this nanotechnology could be applied in our daily lives. And I would look for examples of how nanotechnology could be applied" (Sic).

Knowledge can be an anchor for new, more in-depth concepts on a given subject. Although much debated in multimedia, nanotechnology may not have aroused students' curiosity to understand it better. But with a short video/documentary, it's possible to motivate them to find out more about the subject. Thus, the undergraduate student realised that in order to correlate a piece of content from the chemistry curriculum with nanotechnology, it is necessary to have a prior activity that will lead to more in-depth discussions later on.

"In this case, a more in-depth approach to the content could be taken with the students. For example: 1) Nanotechnology in the chemical industry; 2) Technological processes involving nanotechnology; 3) The impact of nanotechnology on society and the world" (Sic).

In the statement above, the graduate student in question proposes integrating nanotechnology into themes that are consolidated as the foundations of scientific and technological development. Thus, he understands that it is necessary to discuss the subject in the chemical industry, in technological processes and, finally, in discussions about the possible impacts of this technology, which is still incompletely understood.

"Various experimental activities can be tackled with this content. For example, when working on the experimental activity of dye adsorption, it is necessary to use some materials that are capable of adsorbing the dye, including bentonite clay. When carrying out the experiment, the students will observe the rapid adsorption, but they will also ask themselves why some materials are able to adsorb in large quantities and others are not. In this context, the teacher should explain the structure of the material by showing its internal structure and correlating this with nanotechnology so that the *students have*

a better idea of this subject" (Sic).

The above statement shows that the experimental activity can be used to motivate concern based on observing the experiment and then problematising the situation presented to the student. Then come the developments in which the content mentioned by the graduate can be correlated with nanotechnology.

All the methodologies presented to science teaching are responses to the impasses suffered in the face of teaching and learning problems. Lack of stimulus on the part of students to study science, lack of motivation on the part of teachers to innovate their pedagogical practices and inadequate infrastructure in most public schools. However, investigative activities add to the methodologies for teaching and learning science, since:

> When we talk about Science Teaching by Inquiry, we mean to suggest alternative images of science lessons, different from those that have been more common in schools, such as the teacher making notes on the board, followed by explanations and the students taking notes and listening to him or her talk about a particular content topic (MUNFORD and LIMA, 2007, p. 04).

The aim is to give new meaning to the process of acquiring scientific knowledge, demonstrating that science is less abstract than it seems. When students are faced with a research problem, they don't have to play the boring passive role that they are used to, but rather the exciting research activity carried out by researchers and scientists.

4.5 THE MOTIVATION NEEDED TO INCLUDE NANOTECHNOLOGY IN CHEMISTRY LESSONS

Analysing the Course Pedagogical Project (PPC) of the IFRN Chemistry Degree, from which a class was chosen for this case study, it was found that the current curricular proposal, instituted in 2012, complies with the legal determinations contained in the LDB (1996), in CNE/CO opinions 09/2001, 27/2001 and 28/2001, in CNE/CP Resolutions 01/2002 and 02/2002, in the National Curricular Guidelines for Chemistry Degree Courses (CNE/CES Opinion 1.303/2001 and CNE/CES Resolution 08/2002), as well as the IFRN's Political-Pedagogical Project.

It was also found that the course is organised into credit-based subjects, with 2,220 hours of teaching training, 184 hours of curricular seminars and 1,000 hours of professional practice, totalling 3,404 hours. With regard to professional practice, this is seen as the moment when theory and practice are integrated in a process that provides for the formation of professional identity. It is divided as follows: practice as a curricular component (400 hours), teaching internship (400 hours) and academic-scientific-cultural activities (200 hours).

The curriculum is organised around core areas of knowledge, favouring interdisciplinary and contextualised teacher training; strengthening the recognition of the need to train professionals in the area, with regard to scientific and cultural knowledge, the ethical and aesthetic values inherent in learning processes; the socialisation and construction of knowledge acquired in different systemic worldviews.

The four articulating nuclei are: the fundamental, which comprises certain scientific knowledge that is indispensable for the academic and scientific performance and development of undergraduates; the specific, which comprises the fundamental scientific knowledge inherent in the training of chemistry teachers who are able to work in basic education; the epistemological, comprising knowledge about the historical, philosophical, methodological, scientific and linguistic foundations applied to the appropriation of specific knowledge; and finally, the didactic pedagogical, which underpins the teaching knowledge necessary for professional performance in basic education, such as didactic knowledge, cognitive learning processes, understanding of the processes of organisation and management of pedagogical work, among others.

Figure 14 illustrates the organisation and integration between the articulating nuclei of the chemistry degree programme:

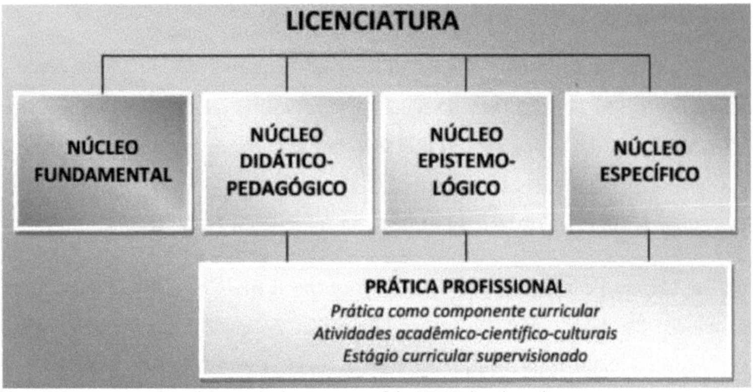

Figure 14: Representation of the curricular structure of the IFRN chemistry degree course
Source: Pedagogical Project for Degree Courses (IFRN, 2012)

Therefore, assuming that the initial training of chemistry teachers at the IFRN is fairly cohesive in terms of content, didactic-pedagogical training and articulated with critical-reflective teaching practice, it is hoped that the undergraduates will be able to integrate the conceptual bases of nanotechnology into their classes as a way of facilitating the understanding of new technologies.

According to Maldaner and Zanon (2013), students live surrounded by technology that in some way derives from science, in its multiple fields, produced by modern man, and within this technology there is a range of knowledge. Therefore, the school must not lose focus on the specificities of the ways in which this knowledge is learnt, since to neglect the broad and comprehensive scientific education of students is to backtrack on the advances in science teaching that have been achieved amid many setbacks.

Consequently, when it comes to the initial training of teachers, it is important to understand how they behave when faced with issues related to the production of new technologies resulting from scientific advances. Nanotechnology, reported in this case study, is just one example, but various themes could be raised to deepen discussions in initial teacher training in the area of science.

Some questions in this regard can be asked: How do future teachers intend to work with themes such as nanotechnology during their professional lives? Do they see it as important for their students' education in understanding the world around them? Are they being prepared for this? What are the challenges and possibilities of teaching nanotechnology to promote interdisciplinarity and contextualisation?

With this in mind, the class surveyed was asked about the importance of including the teaching of nanotechnology during their internship. Paradoxically, even with the didactic activities of selecting content that relates to nanotechnology and is part of the planning required for the internship, it was realised that 75% of the undergraduates do not intend to work with nanotechnology during their internship.

The internship, as seen above, is the moment when impossibilities, challenges, fear of making mistakes and insecurity are confronted with a sense of challenge, where it is possible to lay the foundations for the training of future teachers. It's a privileged moment to take a risk, to dare to break away from the traditional sequence of teaching seen by many students as boring and discouraging, always having a guiding master at your disposal, the figure of a more experienced teacher, who shares your fears and guides you along more coherent paths.

The main reasons cited by the undergraduates for not working on nanotechnology are highlighted below:

"It would be necessary to have a better grasp of this subject, so that I could work on it in the (Internship ((Sic).

"I'm not a big fan of this subject, and I haven't mastered it very well and I feel insecure about working with it in the classroom. I know almost nothing about it" (Sic).

"It's a subject that I don't have a full grasp of, and I also "believe" that the content of the internship is not appropriate for the subject in question" (Sic).

In the above statements by the undergraduates, there was some distance between their conceptual mastery of nanotechnology and the possibility of linking it to the content worked on during the internship. It's worth pointing out that these same students took part in the activity to select content that they could work on during the internship and that would be related to the topic discussed. However, the insecurity of tackling a new subject is a major concern for aspiring teachers. If there is no initial training of these undergraduates to deepen their discussions on the use of new themes, there is a risk of training professionals who are not reflective about their role as trainers.

In addition to the "mastery of the subject" factor, some undergraduates said that they didn't see a direct relationship between chemistry and nanotechnology, or even related it to possible difficulties their students might have. It's important to emphasise that both chemistry and nanotechnology have the same object of study, which is atoms, molecules and their transformations, with the difference being that nanotechnology has a practical application of the knowledge derived from chemistry.

Some also cited the impossibility of tackling the subject due to the school's lack of infrastructure, such as the absence of laboratories with equipment and reagents available and the large number of students. As you can see, in the experimental approaches worked on with the undergraduates, it was made very clear that all experiments developed in an investigative way can be worked on in the classroom, and that a laboratory environment is not necessary.

"The school environment where I work during my internship is not favourable, because the class is quite large and very noisy, it's difficult to get along" (Sic).

"I've already finished my internship and the school where I intend to do my internship II has a laboratory, but it doesn't have adequate (Stc)ture" (Sic).

Lastly, among the undergraduates who said they didn't work on the subject during their internship, there were those who thought the subject was interesting and important, but didn't have an affinity for the subject or didn't see basic education students with the capacity for practical and reflective analysis to understand nanotechnology concepts. Despite this, there is an understanding among educators that the subject's education should be comprehensive and that it should provide moments of difficulty or obstacles that will push them to progress in knowledge and understanding. This critical analysis of their students' education cannot be forgotten by teachers. They are seen as encouragers and motivators in the search for new knowledge.

"The topic is interesting, but I have no affinity with it, and it would be important to find out what the students think" (Sic).

"The topic is relevant (important), however the construction of knowledge must start from a practical-reflexive analysis of the reality around the students, so the topic proposed here still contains (a) disagreement" (Sic).

Despite a certain disparity in the survey's final diagnosis, 25 per cent of the undergraduates considered the possibility of working on nanotechnology during their internship. They saw the subject as a favourable source for inserting new content or for linking knowledge of chemistry with the production of new technologies.

The following are some of the graduate students' comments on their motivation to teach nanotechnology.

"I intend to work on the topic, but I don't know how to apply it, since the content I'm working on with the class all involves calculus, such as calculus content and formulas
((Stchiometrical (Sic) calculations.

"Within the contents of the 3rd year of secondary school, giving credibility to the development of research in the area of nanotechnology, being introductory content to organic chemistry" (Sic).

In the first speech, interest is perceived, but this is countered by some difficulties inherent in the content worked on during the internship. The idea that not all the content worked on during the internship is directly related to nanotechnology is conceivable, given that it has become clear that the subject cannot be considered a structuring theme for all chemistry concepts and skills.

In the second speech, the teaching of nanotechnology is understood as an introductory subject to organic chemistry. The study of carbon, its allotropic forms and the discovery of carbon nanotubes can lead to various discussions and subjects for research, seminars and debates in the classroom. This knowledge can be brought to bear on the following contents in a cross-curricular way.

Furthermore, in order to confirm the importance and effective possibility of introducing nanotechnology into chemistry lessons, we present a brief analysis of the internship report of a graduate student who worked on the subject during her internship. In around five lessons, the topic was covered and recalled in the study of the history of chemistry and the evolution of atomic models. The lesson plans can be found in Annex C.

The subject of the lesson taught by the graduate trainee was Nanotechnology and the history of chemistry, where the aim was to highlight the importance of chemistry as a science that enables the creation of new technologies based on the study of matter. The choice of topic on the historicity of chemistry and technological advances reflects the current conception of science teaching, which advocates a critical view based on science as a human construction, on the historical aspects of its production and inserted in cultural, socio-economic and political contexts (PORTO, 2010).

According to Schulz (2009), science teaching that does not stray from its historicity and does not lose its interdisciplinary focus must be based on the possibility of restructuring teaching and research institutions.

According to Toma (2005, p.09) "the issue of nanotechnology must also be included in the education scenario, since the main players in this new era will be young students, who will need to be prepared to enter this labour market". In this way, science teaching cannot neglect the production of new technologies in its curricula, considering their historical aspects and the interdisciplinarity found in nanotechnology. The school thus becomes a place for thinking about the direction of the new technologies that are emerging.

As for the methodology adopted by the graduate student, it was based on the use of videos and discussion of the students' previous conceptions, as well as a systemic approach to the main concepts for developing nanotechnology skills.

*"**METHODOLOGYR** Playing a video on nanotechnology, followed by a discussion on the subject. Then assimilation with the history of chemistry, relating it to historical points in chemistry. Finally, a video on alchemy.*

(after) the discussion (about the (idco "(Sic).

As can be seen, the video medium was widely used to improve understanding of the topics covered during the lessons, with the aim of making the concepts inherent in nanotechnology easier to understand through the dynamics of the images and audiovisual illustrations. This dynamic of introducing new themes through different methodologies provides a more interesting perspective for the students, who begin to empathise more with the relationship between teaching chemistry and its context.

According to the graduate student's observations, the topic of nanotechnology was well accepted by the students:

***"Students' reaction to the lesson:** The students showed interest in the subject, since nanotechnology was assimilated with their computers and mobile phones. They paid attention and took part in the lesson" (Sic).*

Therefore, the topic of nanotechnology was approached in chemistry classes in a simple way, demonstrating that it is possible to work on the evolution of modern science, highlighting the contributions of chemistry to its development. Furthermore, it is hoped that this teaching will be based on the aspects of critical reflection on technological advances by the school, so that the subjects have a broad scientific education.

CHAPTER 5

FINAL CONSIDERATIONS

The subject of nanotechnology is very topical and contextualises the evolution of modern science to make social life more comfortable with all the possibilities of its applications, such as in medicine, materials engineering and energy production, among others. This subject needs to be discussed more incisively in basic education, as students' scientific training is essential for understanding natural phenomena and also for technological development.

In this way, the teaching of chemistry is seen as a great opportunity to develop this theme in the scientific education of primary school students, since both nanotechnology and chemistry are based on the study of matter and its manipulation for specific purposes. However, the subject of nanotechnology has not been widely covered in the classroom, and this is partly due to a lack of encouragement on the part of initial training courses for chemistry teachers.

This research showed that it is possible to work on nanotechnology during initial training in a contextualised way, using methodological tools and strategies that are already known in chemistry teaching. The subject was approached during the orientation course for the regency internship with a view to motivating the research subjects to work on the concepts covered and thus diversify their teaching practices even during the internship.

As for how to approach nanotechnology, some methodological strategies were proposed for teaching nanotechnology in secondary schools that are different from the few approaches seen so far, namely internet research on the subject or reading and debates on nanotechnology. In this way, the following approaches were proposed: illustrative materials on nanotechnology, investigative experimental activities that enable the acquisition of concepts and skills in nanotechnology and the systematisation of chemistry content with nanotechnology through concept maps.

The challenges for the initial training of chemistry teachers certainly include the need to educate undergraduates about new topics that discuss the development of science and technology. There is a need for more comprehensive curricula that include the teaching of nanotechnology, for example, in their didactic-pedagogical proposals in a more effective way and not simply be left on the sidelines of the training process, which when used is in a way that is disconnected from their training.

With regard to the effective use of the subject by the undergraduates surveyed, it was possible to observe that the majority were interested in the subject, but didn't feel safe working with it during their internship. The reasons for this could lie in the insecurity common to trainees, their lack of mastery of the subject because they don't know it well or the resistance of the subject teachers to breaking the sequence of content in the curriculum adopted by the school.

It is also important to encourage and accompany undergraduates to develop the teaching of topics that seem challenging to them, because the figure of a teacher who is a mentor during teaching practice can spread the feeling of fear and refusal to work on topics such as nanotechnology.

These conjectures are possibilities that should be worked on to improve research in chemistry teaching, especially in initial teacher training. It is known that science education in Brazil has a short tradition, but over the last few years several studies have pointed to the need for specific teacher training in the area and for research related to the teaching of each subject. The shortfall in specific science teacher training is being met through public policies such as the creation and internalisation of Federal Institutes in Brazilian states. However, this growth cannot be seen in isolation from the many variables involved in the process of educating students in science, but rather to promote discussion about improvements in initial teacher training with positive repercussions on basic education.

The methodological approach used in this case study is purely qualitative and favours the speeches and conceptions of the research subjects, as this highlights their uniqueness and shows the importance of their role in teaching research. The activities developed during the meetings presented the foundations of the research-based teaching strategy, which is always concerned with critical reflection rather than ready-made answers. As the aim was for the undergraduates to feel encouraged to work on nanotechnology in their internships, several moments of reflection and discussion were spent looking for ways of transposing the activities into primary school classrooms.

BIBLIOGRAPHICAL REFERENCES

ALFONSO, A. B. **Current situation of dissemination and training in nanoscience and nanotechnology in Brazil**. Revista Interdisciplinaria en Nanociencias yNanotecnología. Coyoacán, Mexico: Universidad Nacional Autónoma de México, v.4, 2011.

ALMEIDA, P. C. A. e BIAJONE, J. **Saberes docentes e formação inicial de professores**: implicações e desafios para as propostas de formação. Educação e Pesquisa (USP), v. 33, p. 281-296, 2007.

ALVES, O. L. **Nanotecnologia, nanociência e nanomateriais**: quando a distância entre presente e futuro não é apenas questão de tempo. Parcerias Estratégicas (Impresso), Brasília, v. 18, p. 23-40, 2004.

ABDI, Brazilian Agency for Industrial Development. **Nanotechnology Prospective Study**: 2008-2025. Available at: www.desenvolvimento.gov.br/arquivos/dwnl_1266607820.pdfAcesso: 01/05/2013.

ANDRADE, J. B. et al . **Chemistry in Brazil**: Prospects and Needs for the Next Decade. Química Nova (Impresso), v. 28, p. 7-10, 2005.

ARROIO, A.; GIORDAN, M. **O Vídeo Educativo**: aspectos da organização do ensino. Química Nova na Escola, v. 24, p. 8-12, 2006.

AVELAR, F. F. et al. **The use of piassava fibres (Attalea funifera) in the preparation of activated carbon**. Bioresource Technology, 2010, p. 4639-4645.
Available at: http://www.sciencedirect.com/science/article/pii/S0960852410001896

AZEVEDO, R. B. de. **Electron microscopy**. DURÁN, N.; MATTOSO, L. H. C. E MORAIS, P. C. de. (Organisers). In: Nanotechnology: introduction, preparation and characterisation of nanomaterials and application examples. São Paulo: Artliber Editora, 2006.

BACHELARD, G. **The Formation of the Scientific Spirit:** Contribution to a Psychoanalysis of Knowledge. Trad. Estela dos Santos Abreu. - Rio de Janeiro: Contraponto, 1996.

BASSOTTO, G. V. **Nanotecnologia: uma investigação fundamentada na educação pela pesquisa se refletindo na formação de professores e no ensino de química**. Rio Grande do Sul, 129 p., 2011. Dissertation (Master's) - Pontifical Catholic University of Rio Grande do Sul.

BELLUCCO, A. and CARVALHO, A. M. P. de. **A proposal for an investigative teaching sequence on quantity of movement, its conservation and Newton's laws.** Caderno Brasileiro de Ensino de Física, v. 31, p. 30-59, 2014.

BERNAL, J. D. **Science in History**. Lisbon: Livros Horizonte, 1969. Translation: Antônio Neves Pedro, 1st volume, movement collection.

BOGDAN, R. and BIKLEN, S. **Qualitative research in education:** an introduction to theory and methods. Porto: Porto Editora, 1994.

BORSCHIVER, S. et al. **Patenting in Nanotechnology**: A Study of the Nanostructured Polymeric Materials Sector. Polímeros: Ciência e Tecnologia, vol. 15, n° 4, p. 245-248, 2005 (Department of Organic Processes, School of Chemistry, UFRJ).

BRASIL, **Parâmetros Curriculares Nacionais do Ensino Médio (PCNEM), Part III - Natural Sciences, Mathematics and their Technologies**, 2000.
Available at: *portal.mec.gov.br/seb/arquivos/pdf/ciencian.pdf* Accessed on: 18/05/2013.

BRAZIL. **Curricular Guidelines for Secondary Education - Natural Sciences, Mathematics and their Technologies**. 2006. Brasilia. Available at: <http://portal.mec.gov.br/seb/arquivos/pdf/book_volume_02_internet.pdf> Accessed on: 06 Jan. 2015.

BRAZIL. **National Curriculum Guidelines for Basic Education**. 2013. Available at: <http://portal.mec.gov.br/index.php?option=com_content&view=article&id=12663&Itemid=1152> Accessed on: 06 Jan. 2015.

BRAZIL. **National Curriculum Parameters for Secondary Education**: Part III - Natural Sciences, Mathematics and their Technologies. Brasilia, 1998.

CARVALHO, A. M. P. de. (Org.) **Teaching Science by Inquiry**: conditions for implementation in the classroom. São Paulo: Cengage Learning, 2013.

CARVALHO, A. M. P. de, e GIL-PÉREZ, D. **Formação de professores de ciências**: tendências e inovações. 10. ed.- São Paulo: Cortez, 2011.

CHASSOT, A. I. **Is science masculine? Yes, it is!** 7. ed. São Leopoldo: UNISINOS, 2015, *Aldus* collection, 134p .

DIMER, F. A.; FRIEDRICH, R. B.; BECK, R. C. R.; GUTERRES, S. S. and POHLMANN, A. R. **Impacts of nanotechnology on health**: drug production. Química Nova, Vol. 36, No. 10, 1520-1526, 2013.

ELLWANGER, A. L. et al. **Nanoscience teaching through learning objects**. RENOTE. New Technologies in Education Journal, v. 10, p. 01-10, 2012. Available at: *seer.ufrgs.br/rendte/article/dowdoad/30884/19237*. Accessed on: 28 March 2013.

FERNANDES, F. M. **Synthesis of orientated carbon nanotubes and application in the production of AFM tips**. Dissertation (Master's Degree) - University of São Paulo. Institute of Physics - Department of Applied Physics.

FERNANDES, M. F. M.; FILGUEIRAS, C. A. L.. **An overview of nanotechnology in Brazil (and its macro-challenges)**. Química Nova (Impresso), v. 31, p. 2205-2213, 2008.

FERREIRA, L. H.; HARTWIG, D. R. and OLIVEIRA, R. C. de. **Experimental Chemistry Teaching**: A Contextualised Investigative Approach. Química Nova na Escola, v. 32, n° 2, 2009

FERRÉS, J. **Cómo integrar el vídeo en la escuela**. Barcelona: Ceac, 1988.

FURIÓ, C.; GIL-PÉREZ, D. La didáctica de las ciências em la formación inicial del professororado: uma orientación y um programa teóricamente fundamentados. Ensenanza de las Ciencias, v.7, n. 3, p. 257-265, 1989.

GIL, A. C. **Métodos e técnicas de pesquisa social**. 6. ed. - São Paulo: Atlas, 2009.

GIL, A. C. **Como elaborar projetos de pesquisa**. 5. ed. - São Paulo: Atlas, 2010.

GOUVEIA, C.A. ; GARVIL, M. P. ; ARANTES, D. E. . **Nanotechnology in Cosmetics**. In: XVII Annual Science Meeting, 2013, Uberlândia. Proceedings of the 17th Annual Science Meeting. Uberlandia: erac, 2013. v. 3.

HERBST, M. H. ; MACEDO, M. I. F. ; ROCCO, A. M. . **Carbon Nanotube Technology: Trends and Perspectives in a Multidisciplinary Field**. Química Nova (Impresso), v. 27, n.6, p. 986-992, 2004.

HODSON, D. **Phylosophy of Science.** Science and Science Educations. Studies in Science Eduction, s. l., v. 12, 1985, p.8.

JENSEN, T. F. **Nanotecnologia**: maravilhas e incertezas no universo da química. São Paulo: Fundacentro, p. 24, 2010 (Nanotecnologia em quadrinhos; n. 2).

LACAVA, Zulmira G. M. **Biomedical applications of magnetic nanoparticles**. In: DURAN, N.; MATTOSO, L. H. C.; MORAIS, P. C. de. (Org.). NANOTECHNOLOGY: Introduction, preparation and characterisation of nanomaterials and application examples. 01ed.São Paulo: Artliber Editora, 2006, p. 175-181.

LAMPTON, C. **Having Fun with Nanotechnology.** Rio de Janeiro: Berkeley, 1994. Translation by

Amir Kurban.

LIBÂNEO, J. C. **Goodbye teacher, goodbye teacher? :** new educational demands and the teaching profession. 13th ed. - São Paulo: Cortez, 2011.

LIMA, M. C. A.; ALMEIDA, Maria José P. M. **Articulation of texts on nanoscience and nanotechnology for the initial training of physics teachers**. Revista Brasileira de Ensino de Física (Impresso), v. 34, p. 4401-4409, 2012.

LÔBO, S. F. **The teaching of chemistry and the formation of the chemical educator, under a Bachelardian perspective**. Ciência & Educação, v. 14, n. 1, p. 89-100, 2008.
Available: ...www.scielo.br/scielo.php?pid=S151673132008000100006&script
Accessed: 15/04/2013.

LUDKE, M. and ANDRÉ, M. E. D. A. de. **Research in education:** qualitative approaches. São Paulo: EPU, 1986.

LUDKEN, M. **Discussion of the work of Robert E. Stake case study in educational research and evaluation**. Education and Selection. São Paulo, Carlos Chagas Foundation, n° 7, 1983.

MALDANER, O. A., and SANTOS, L. P. dos. **Teaching Chemistry in Focus**. Ijuí: Ed. Unijuí, 2010.

MEDEIROS, E. S. de; MATTOSO, L. H. C. **Nanotechnology**. In: Nelson Eduardo Durán Caballero; Luiz Henrique Capparelli Mattoso; Paulo César de Morais. (Org.). Nanotechnology: Introduction to the preparation and characterisation of nanomaterials and application examples. São Paulo: Artliber Editora, 2006.

MOREIRA, M. A. **Concept maps and meaningful learning.** São Paulo: Centauro, 2010.

MUNFORD, D. and LIMA, M. E. C. C. E. **Teaching science through enquiry**: what do we agree on? Essay. Research in Science Education, v. 9, p. 20, 2007.

NAOE, A. **Nanoart collaborates in teaching and publicising the field.** USP News Agency (USP Online), 2015. Available at:www.usp.br/agen/?p=199829 . Accessed on: 30/04/2015.

NUNEZ, I. B. and RIBEIRO, R. P.. **Meaningful learning and the teaching of natural sciences.** In: NUNEZ, I. B.; RAMALHO, B. L (ORGS). Fundamentals of teaching and learning natural sciences and maths: the new secondary education. Porto Alegre: Sulina, 2004. 300p.

OLIVEIRA, L. H. M. and CARVALHO, R. S. **A look at the History of Chemistry in Brazil**. Ponto de Vista Magazine, v. 3, p. 27-37, 2006.

PAULINI DE JESUS, I. and HIGA I. **Nanotechnology and secondary education**: a literature review on teaching proposals. In: IV National Symposium on Science and Technology Teaching, Ponta Grossa, 2014.

PASCHOALINO, M. P.; MARCONE, G. P. S.; JARDIM, W. F. **Nanomaterials and the environmental issue** . Química Nova (Impresso), v. 33, p. 421-430, 2010.

POLETTE, F. **Magic Bullet**. Science Blogs: science, culture and politics, 2012. Available at: http://scienceblogs.com.br/balamagica/category/imagens-artisticas/ Accessed on: 30/04/2015.

PORTO, P. A. **History and Philosophy of Science in Chemistry Teaching:** In search of today's educational objectives. In: SANTOS, W. L. P. e MALDANER, O. A. (Org.) Ensino de Química em foco. Ijuí: Ed. Unijuí, 2010.

QUINA, F. H. **Nanotechnology and the Environment**: Perspectives and Risks. Química Nova, v 27, n. 6, p. 1028-1029, 2004.

REBELLO, G. A. F. et al. **Nanotechnology, a theme for secondary education using the CTSA approach.** Química Nova Escola, v. 34, p. 3-9, 2012.

REIS, M. S. A. **Comic books as a teaching resource in science education.**
Ensino e Re-Vista, Uberlândia, v. 9, n.1, p. 105-114, 2001.

RIBEIRO, R. P.; NUNEZ, I. B.. **Thinking about meaningful learning:** from concept maps to concept networks. *In:* NUNEZ, I. B; RAMALHO, B. L. (Org.). Fundamentos do ensino-aprendizagem das ciências naturais e da matemática: O novo ensino médio. Porto Alegre: Sulina, 2004

RIBOLDI, B. M. **Nanotechnology**: fundamentals and applications. 2009. 5th UNESP university extension congress. Available:*www.proex-unesp.com.br/6congresso/* Accessed on: 10/07/2013.

ROSA, P. R. S. **The use of audiovisual resources and science teaching**. Caderno Catarinense de Ensino de Física, Florianópolis, v. 17, n.1, p. 33-49, 2000.

ROSSI-BERGMANN, B. **Nanotechnology**: health beyond technological determinism. Ciência e Cultura (SBPC), v. 60, p. 54-57, 2008.

SÁ, L. P. and QUEIROZ, S. L. **Case studies in chemistry teaching**. Campinas, SP: Editora Átomo, 2010.

SAGAN, C. **O mundo assombrado pelos demônios**: a ciência vista como uma vela no escuro. São Paulo: Companhia das letras, 1996.

SALES, P. F. de. et al. **Production, Characterisation and Application of Activated Carbon Obtained from Corn Cob**: The Search for the Reuse of an Agroindustrial Waste. Revista Virtual Química, 2015, vol. 7, n. (4).

SANNOMIYA, M. ; HONÓRIO, K. M. ; PEREIRA, F. D. . **Nanotechnology:** Development of Teaching Materials for an Approach in Elementary and Secondary Schools. Química Nova na Escola, v. 32, n. 2, 2010.

SCHNETZLER, R. P. **Notes on the History of Chemistry Teaching in Brazil.** In: MALDANER, O. A., and SANTOS, L. P. dos. Teaching Chemistry in Focus. Ijuí: Ed. Unijuí, 2010.

Research in Chemistry Teaching in Brazil: Achievements and Perspectives. Química Nova, São Paulo, v. 25, p. 14-24, 2002.

SCHULZ, P. A. **Nanomaterials and the interface between nanotechnology and the environment**. Health Surveillance in Debate: Society, Science & Technology, v. 1, p. 53-58, 2013.

. **The crossroads of nanotechnology**: innovation, technology and risks. Rio de Janeiro: Viera & Lent, 2009.

. **Low-cost nanoscience at home and at school**. Physics at School, v. 8, n. 1. 2007.

SCHUCH, F; BEVILAQUA, R. ; FAGAN, S. B. . **Electronic and Structural Properties of Carbon Nanotubes and Application as Drug Carriers**. Scientia. Natural and Technological Sciences Series, v. 8, p. 95-105, 2007.

SCHNETZLER, R. P. e ARAGÃO, R. M. R. **Importância, sentido e contribuições de pesquisas para o Ensino de Química**. Química Nova na Escola. São Paulo. n.1, p.27-31, 1995.
SCLIAR-CABRAL, L. **Goals for teacher training:** priorities. Acts of Research in Education (FURB), v. 2, p. 197-206, 2007

SILVA, S. L. A.; VIANA, M. M.; MOHALLEM, N. D. S.. **After all, what is Nanoscience and Nanotechnology? An Approach for Secondary Schools**. Química Nova na Escola (Impresso), v. 31, p. 1-7, 2009.

SILVA, C. S. da, and OLIVEIRA, L. A. A. de. **Initial training for chemistry teachers**: specific and pedagogical training. In: Roberto Nardi. (Teaching Science and Maths I: themes on teacher training. 1ed. São Paulo: Cultura Acadêmica, v. 1, 2010.

SILVA, C. G. da. **Nanotechnology:** the national challenge. Strategic Partnerships, n. 18, Brasília, 2004.

SILVA , J. L. da. et al. **The Use of Teaching Videos in High School Chemistry Classes for a Historical and Contextualised Approach to the Glass Theme**. Química Nova na Escola, Vol. 34, n. 4, p. 189-200, 2012.

SILVA, M. B. **Nanotechnology**: New Ethical Issues for Brazil, Legal and Social Dimensions in an Interdisciplinary Approach. Cadernos de Pesquisa Interdisciplinar em Ciências Humanas (UFSC), v. 46, p. 1-17, 2003.

SIQUEIRA-BATISTA, R. et al. **Nanotechnology and science teaching in the light of the CTS approach**: a trip to Lilliput. Revista Ciências & Idéias, v. 1, p. 76-86, 2009.

STAKE, R. **Case studies in educational research and evaluation.** Education and Selection. São Paulo, Carlos Chagas Foundation, No. 7, 1983.

STAKE, R. **Qualitative/naturalistic research - epistemological problems.** Education and Selection. São Paulo, Carlos Chagas Foundation, No. 7, 1983.

TASCA, R. A. et al. **Developing Nanotechnology Skills and Concepts in High School through a Didactic Experiment Involving the Preparation and Application of Superparamagnetic Nanoparticles**. Química Nova na Escola, v. 37, n° 3, p.236-240, 2015.

TEDESCO, A. C. **Use of Nanostructured Devices in Cancer Treatment**. In: Rede Cooperativa de Nanotecnologia. (Org.). Nanotecnologia (in press). São Paulo: Artliber Editora LDTA, 2006, p. 183-196.

TEIXEIRA, F. M. and SOBRAL, A. C. M. **Prior knowledge**: investigating how it is used by science teachers in the early grades of primary school. Ciência e Educação (Unesp. impresso), v. 16, p. 667-677, 2010.

TEIXEIRA, F. S. and SALVADORI, M. C. **Scanning electron microscope as a tool for micro and**

nanotechnology. Mecatrônica Fácil, v. 31, p. 38 - 41, 2007.

TOMA, H. E. **The Nanotechnology of Molecules**. Química Nova na Escola, Brazil, v. 21, p.3- 9, 2005.

_______. **Interfaces and organisation of research in Brazil**: From Chemistry to Nanotechnology. Química Nova, São Paulo, SP - Brazil, v. 28, n. Suplemento, p. S48-S51, 2005.

TRINDADE, J. O.; HARTWIG, D. R.. **Combined use of concept maps and diversified teaching strategies:** an initial analysis of chemical bonds. Química Nova na Escola (Impresso), v. 34, p. 83-91, 2012.

TRÓPIA, G. **Historical paths of teaching science through investigative activities.** Ensaio: Pesquisa em Educação em Ciências (Online), v. 13, p. 121-137, 2011.

VERGUEIRA, W. et al. **How to use comic strips in the classroom**. 4. ed., 2ª reprint - São Paulo: Contexto, 2014.

VIANNA, J. F.; AYDOS, M. C. R.; SALGADO, O. S. **Night Chemistry Degree Course**: A decade of experience at UFMS. Química Nova, v. 20, n.2, p. 213-218, 1997.

YAMAMURA, A. P. G. **Application of nanotechnology in the environment**: magnetic biosorbent for uranium removal. São Paulo, 2009. Master's dissertation (Nuclear and Energy Research Institute, USP).

YAMAURA, M. et al. **Preparation and characterisation of (3- aminopropyl)triethoxysilane-coated magnetic nanoparticles.** J. Magnetism and Magnetic Materials, n. 279, p. 210-217, 2004.

YIN, R. K. **Estudo de caso**: planejamento e métodos. 2. ed. Porto Alegre: Bookman, 2001.

ZANON, L. B. et al. **Natural Sciences, Mathematics and their Technologies**: Knowledge of Chemistry. In: Ministry of Education, Secretariat of Basic Education.
(Curriculum Guidelines for Secondary Education: Natural Sciences, Mathematics and their Technologies. Brasília: Ministry of Education, 2006, v. 2, p. 99-137.
Available at: http://portal.mec.gov.br/seb/arquivos/pdf/09Quimica.pdf Accessed: 02/10/2015.

ZOMPERO, A. F. and LABURÚ, C. E. **Investigative activities in science teaching**: historical aspects and different approaches. Ensaio: Pesquisa em Educação em Ciências (Impresso), v. 13, p. 67-80, 2011.

ZUCCO, C. **Chemistry graduation: a new chemist for a new era**. Química Nova, São Paulo, v. 28, Suplemento, p. S11-S13, 2005.

MIX
Papier aus verantwortungsvollen Quellen
Paper from responsible sources
FSC® C105338

Printed by Books on Demand GmbH, Norderstedt / Germany